Shale Oil Production Processes

James G. Speight, PhD, DSc
CD&W Inc., Laramie, Wyoming, USA

AMSTERDAM • BOSTON • HEIDELBERG • LONDON
NEW YORK • OXFORD • PARIS • SAN DIEGO
SAN FRANCISCO • SINGAPORE • SYDNEY • TOKYO
ELSEVIER   Gulf Professional Publishing is an imprint of Elsevier

Gulf Professional Publishing is an imprint of Elsevier
The Boulevard, Langford Lane, Kidlington, Oxford, OX5 1GB, UK
225 Wyman Street, Waltham, MA 02451, USA

First published 2012

British Library Cataloguing in Publication Data
A catalogue record for this book is available from the British Library

Library of Congress Cataloging-in-Publication Data
A catalog record for this book is available from the Library of Congress

ISBN: 978-0-12-401721-4

For information on all Gulf Professional Publishing publications
visit our website at *store.elsevier.com*

This book has been manufactured using Print On Demand technology. Each copy is produced to order and is limited to black ink. The online version of this book will show color figures where appropriate.

Working together to grow
libraries in developing countries

www.elsevier.com | www.bookaid.org | www.sabre.org

ELSEVIER   BOOK AID
            International   Sabre Foundation

TABLE OF CONTENTS

Oil shale is a fine-grained sedimentary rock containing organic matter (commonly called *kerogen*) that yields substantial amounts of oil and combustible gas upon destructive distillation. Most of the organic matter is insoluble in ordinary organic solvents; therefore, it must be decomposed by heating to release such materials. Underlying most definitions of oil shale is its potential for the economic recovery of energy, including shale oil, combustible gas, and a number of by-products. A deposit of oil shale having economic potential is generally one that is at or near enough to the surface to be developed by open-pit or conventional underground mining or by in situ methods.

Oil shale deposits are found in many parts of the world. They range in age from Cambrian to Tertiary and were formed in a variety of marine, continental, and lacustrine depositional environments. The largest known deposit is in the Green River Formation in the western United States with the potential to produce approximately 1.5 trillion U.S. barrels (1.5×10^{12} U.S. bbls) of shale oil.

The total resources of a selected group of oil shale deposits in various countries have the potential to produce 2.8 trillion U.S. barrels (2.8×10^{12} U.S. bbls) of shale oil. These amounts are very conservative because (1) several deposits have not been explored sufficiently to make accurate estimates and (2) some smaller deposits were not included in these numbers.

The properties of shale oil vary with its source and the process by which it is produced. Thus, shale oil can be a difficult feedstock to process. Varying quantities of heteroatoms (nitrogen, oxygen, sulfur, and metal constituents) offer several difficulties that refiners must face if they are to include shale oil as part of the refinery feedstock. In addition, the incompatibility of shale oil with typical petroleum feedstocks may also be an issue.

However, depending on the nature of the upgrading techniques applied, shale oil can be a premium-quality refinery feedstock, comparable and compatible with the best grades of conventional crude oil. In fact, shale oil is considered by some scientific and engineering authorities to be a better source of jet fuel, diesel fuel, and distillate heating oil than it is of gasoline. Although some technical questions remain, the upgrading and refining processes are well-advanced for the production of premium products.

This book deals with the production of shale oil from oil shale and transports the reader through the various aspects of shale oil production, with chapters that deal with (1) origin and properties of oil shale, (2) oil shale resources by country, (3) the chemical and physical nature of kerogen, the precursor to oil shale, (4) mining and reporting oil shale, (5) in situ retorting, (6) refining shale oil, and (7) the environmental aspects of shale oil production. A descriptive Glossary is also included.

Dr. James G. Speight
Laramie, Wyoming,
June 2012

CHAPTER 1

Origin and Properties of Oil Shale

1.1 ORIGIN

Oil shale represents a large and mostly untapped hydrocarbon resource. Like tar sand (*oil sand* in Canada) and coal, oil shale is considered unconventional because oil cannot be produced directly from the resource by sinking a well and pumping. Oil has to be produced thermally from the shale. The organic material contained in the shale is called *kerogen*, a solid material intimately bound within the mineral matrix (Allred, 1982; Baughman, 1978; Lee, 1996; Scouten, 1990; US DOE, 2004a,b,c; Speight, 2007, 2008, 2013).

Oil shale is distributed widely throughout the world with known deposits in every continent. Oil shale ranging from Cambrian to Tertiary in age occurs in many parts of the world (Table 1.1). Deposits range from small occurrences of little or no economic value to those of enormous size that occupy thousands of square miles and contain many billions of barrels of potentially extractable shale oil. However, petroleum-based crude oil is cheaper to produce today than shale oil because of the additional costs of mining and extracting the energy from oil shale. Because of these higher costs, only a few deposits of oil shale are currently being exploited; in China, Brazil, and Estonia. However, with the continuing decline of petroleum supplies, accompanied by increasing costs of petroleum-based products, oil shale presents an opportunity for supplying some of the fossil energy needs of the world in the future (Andrews, 2006; Bartis et al., 2005; Culbertson and Pitman, 1973).

Oil shale is not generally regarded as true shale by geologists nor does it contain appreciable quantities of free oil (Scouten, 1990; Speight, 2008). The fracture resistance of all oil shales varies with the organic content of the individual lamina, and fractures preferentially initiate and propagate along the leaner horizontal laminas of the depositional bed.

Table 1.1. Estimate of Oil Shale Reserves (Tons $\times 10^6$)

Region	Shale Reserves	Kerogen Reserves	Kerogen in Place
Africa	12 373	500	5900
Asia	20 570	1100	–
Australia	32 400	1700	37 000
Europe	54 180	600	12 000
Middle East	35 360	4600	24 000
North America	3 340 000	80 000	140 000
South America	–	400	10 000

Source: World Energy Council, WEC Survey of Energy Resources.
To convert tons to barrels, multiply by 7 indicating approximately 620 billion barrels of known recoverable kerogen, which has been estimated to be capable of producing 2600 billion barrels of shale oil. This compares with 1200 billion barrels of known worldwide petroleum reserves (Source: BP Statistical Review of World Energy, 2006).

Table 1.2. General Classification of Oil Shale

Sedimentary Rocks	Humic coal
Nonorganic	Bitumen-containing
Organic rich	Tar sand (oil sand)
	Oil shale
	Terrestrial
	Cannel coal
	Lacustrine
	Lamosite
	Torbanite
	Marine
	Kukersite
	Marinite
	Tasmanite

Oil shale was deposited in a wide variety of environments, including freshwater to saline ponds and lakes, epicontinental marine basins, and related subtidal shelves as well as shallow ponds or lakes associated with coal-forming peat in limnic and coastal swamp depositional environments. These give rise to different oil shale types (Table 1.2) (Hutton, 1987, 1991), and therefore, it is not surprising that oil shales exhibit a wide range of organic and mineral compositions (Mason, 2006; Ots, 2007; Scouten, 1990; Wang et al., 2009). Most oil shale contains organic matter derived from varied types of marine and lacustrine algae, with some debris from land plants, depending on the depositional environment and sediment sources.

Organic matter in the oil shale is a complex mixture and is derived from the carbon-containing remains of algae, spores, pollen, plant cuticle, corky fragments of herbaceous and woody plants, plant resins, and plant waxes, and other cellular remains of lacustrine, marine, and land plants (Dyni, 2003, 2006; Scouten, 1990). These materials are composed chiefly of carbon, hydrogen, oxygen, nitrogen, and sulfur. Generally, the organic matter is unstructured and is best described as amorphous (*bituminite*)— the origin of which has not been conclusively identified but is theorized to be a mixture of degraded algal or bacterial remains. Other carbon-containing materials such as phosphate and carbonate minerals may also be present, which, although of organic origin, are excluded from the definition of organic matter in oil shale and are considered to be part of the mineral matrix of the oil shale.

Oil shale has often been called *high-mineral coal*, but nothing can be further from reality. Maturation pathways for coal and kerogen are different, and, in fact, the precursors of the organic matter in oil shale and coal also differ (Durand, 1980; Hunt, 1996; Scouten, 1990; Speight, 2013; Tissot and Welte, 1978). Furthermore, the origin of some of the organic matter in oil shale is obscure because of the lack of recognizable biological structures that would help identify the precursor organisms, unlike the recognizable biological structures in coal (Speight, 2013). Such materials may be of (1) bacterial origin, (2) the product of bacterial degradation of algae, (3) other organic matter, or (4) all of the above.

Furthermore, oil shale does not undergo the maturation process that occurs for petroleum and/or coal but produces the material known as *kerogen* (Scouten, 1990). However, there are indications that kerogen may be a by-product of the maturation process. The kerogen residue that remains in the oil shale is formed during maturation and is then ejected from the organic matrix because of its insolubility and relative unreactivity under the maturation conditions (Speight, 2007; Chapter 4). Furthermore, the fact that kerogen, under the high-temperature pyrolysis conditions imposed upon it in the laboratory, forms hydrocarbon distillates (albeit with relatively high amounts of nitrogen) does not guarantee that the kerogen of oil shale is a precursor to petroleum.

The thermal maturity of oil shale refers to the degree to which the organic matter has been altered by geothermal heating. If oil shale is heated to the maximum highest temperature—the actual historical temperature to which the shale has been heated is not known with any degree of accuracy and is typically speculative—as may be the case if the oil shale were deeply buried, the organic matter *may* thermally decompose to form liquids and gas. Under such circumstances, there is highly unfounded speculation (other than high-temperature laboratory experiments) that oil shale sediments can act as the source rocks for petroleum and natural gas.

Moreover, as stated above, the fact that the high-temperature thermal decomposition of kerogen (in the laboratory) gives petroleum-like material is no guarantee that kerogen is or ever was a precursor to petroleum. The implied role of kerogen in petroleum formation is essentially that—implied, but having no conclusive experimental foundation. However, caution is advised in choosing the correct definition of kerogen since there is the distinct possibility that it is one of the by-products of the petroleum generation and maturation processes, and may not be a direct precursor to petroleum.

Petroleum precursors and petroleum are indeed subject to elevated temperatures in the subterranean formations due to the *geothermal gradient*. Although the geothermal gradient varies from place to place, it is generally in the order of 25°C/km to 30°C/km (15°F/1000 feet or 8°C/1000 feet, i.e., 0.015°F per foot of depth or 0.008°C per foot of depth). This leaves a serious question about whether or not the material has been subjected to temperatures greater than 250°C (>480°F).

Such experimental work is interesting insofar as it shows similar molecular moieties in kerogen and petroleum (thereby confirming similar origins for kerogen and petroleum). However, the absence of geological time in the laboratory is not a reason to increase the temperature and it must be remembered that application of high temperatures (>250°C, <480°F) to a reaction not only increases the rate of reaction (thereby making up for the lack of geological time) but can also change the *nature* and the *chemistry* of a reaction. In such a case, the geochemistry is altered. Furthermore, introduction of a pseudo-activation energy in which the activation energy of the kerogen conversion reactions are reduced leaves much to be desired because of the assumption required to develop this pseudo-activation energy equation(s). In fact, not only will the oil window (the oil-producing phase) vary from kerogen-type to kerogen-type, but it is also not valid to use a fixed set of kinetic parameters within each of these groups.

It is claimed that the degree of thermal maturity of an oil shale can be determined in the laboratory by any one of several methods. One method is to observe the changes in the color of the organic matter in samples collected from varied depths—assuming that the organic matter is subjected to geothermal heating (the temperature being a function of depth), the color of the organic matter might be expected to change from a lighter color (at relatively shallow depths) to a darker color (at relatively deep depths). Then, another unknown issue of shifting of the sedimentary strata comes into play.

Suffice it to state that the role played by kerogen in the petroleum maturation process is not fully understood (Durand, 1980; Hunt, 1996; Scouten, 1990; Speight, 2007; Tissot and Welte, 1978). What obviously needs to be addressed more fully in terms of kerogen participation in petroleum generation is the potential to produce petroleum constituents from kerogen by low-temperature processes rather than by processes that involve the use of temperatures greater than 250°C (>480°F) (Burnham and McConaghy, 2006; Speight, 2007).

If such geochemical studies are to be pursued, a thorough investigation is needed to determine the potential for such high temperatures being present during the main phase, or even various phases, of petroleum generation in order to determine whether kerogen is a precursor to petroleum (Speight, 2007).

Finally, much of the work performed on oil shale has referenced the oil shale from the Green River Formation in the western United States. Thus, unless otherwise stated, the shale referenced in the following text is the Green River shale.

1.2 OIL SHALE TYPES

Mixed with a variety of sediments over a lengthy geological time period, shale forms a tough, dense rock ranging in color from light tan to black. Based on its apparent colors, shale may be referred to as *black shale* or *brown shale*. Oil shale has also been given various names in different regions. For example, the Ute Indians, on observing outcroppings burst into flames after being hit by lightning, referred to it as *the rock that burns.*

Thus, it is not surprising that definitions of the types of oil shale can be varied and confusing. It is necessary to qualify the source of the definition and the type of shale that fits within it.

For example, one definition is based on the mineral content of the shale, in which three categories can recognized namely, (1) carbonate-rich oil shale, which contain a high proportion of carbonate minerals (such as calcite and dolomite) and which usually have the organic-rich layers sandwiched between carbonate-rich layers—these shales are hard formations that are resistant to weathering and are difficult to process using mining (*ex situ*); (2) siliceous oil shales, which are usually dark brown or black. They are deficient in carbonate minerals but plentiful in siliceous minerals (such as quartz, feldspar, clay, chert, and opal)— these shales are not as hard and weather-resistant as the carbonate shales and may be better suited for extraction through mining (*ex situ*) methods; and (3) cannel oil shales, which are typically dark brown or black and consist of organic matter that completely encloses other mineral grains—these shales are suitable for extraction through mining (*ex situ*).

However, mineral content aside, it is more common to define oil shale on the basis of their origin and formation as well as the character of their organic content. More specifically, the nomenclature is related to whether or not the shale is of (1) *terrestrial* origin, (2) *marine* origin, or (3) *lacustrine* origin (Hutton, 1987, 1991). This classification reflects differences in the composition of the organic matter and of the distillable products that can be produced from the shale. This classification also reflects the relationship between the organic matter found in the sediment and the environment in which the organic precursors were deposited.

1.2.1 Terrestrial Oil Shale

The precursors to terrestrial oil shale (sometimes referred to as *cannel coal*) were deposited in stagnant, oxygen-depleted waters on land (such as coal-forming swamps and bogs).

Cannel coal is brown to black oil shale composed of resins, spores, waxes, and cutinaceous and corky materials derived from terrestrial vascular plants, together with varied amounts of vitrinite and inertinite. Cannel coals originate in oxygen-deficient ponds or shallow lakes in peat-forming swamps and bogs. This type of shale is usually rich in oil-generating lipid-rich organic matter derived from plant resins, pollen, spores, plant waxes, and the corky tissues of vascular plants. The individual deposits usually are small in size, but they can be of a very high grade.

The latter also holds for lacustrine oil shales. This group of oil shales was deposited in freshwater, brackish, or saline lakes. The size of the organic-rich deposits can be small, or they can occur over tens of thousands of square miles as is the case for the Green River Formation in Colorado, Utah, and Wyoming. The main oil-generating organic compounds found in these deposits are derived from algae or bacteria. In addition, variable amounts of higher plant remains can be present.

1.2.2 Lacustrine Oil Shale
Lacustrine oil shales (lake-bottom-deposited shales) include lipid-rich organic matter derived from algae that lived in freshwater, brackish, or saline lakes.

The lacustrine oil shales of the Green River Formation, which were discussed above, are among the most extensively studied sediments. However, their strongly basic depositional environment is certainly unusual, if not unique. Therefore, it is useful to discuss the characteristics of the organic material in other lacustrine shales.

Lacustrine sequences from the Permian oil shales of Autun (France) and the Devonian bituminous flagstones of Caithness (Scotland) exhibited several series of biomarkers that were prominent in extracts from these shales—hopanes, steranes, and carotenoids. Algal remains were abundant in both shales. Blue-green algae, similar to those that contributed largely to the Green River oil shale kerogen, were found in the Devonian shale, for which a stratified lake environment similar to Green River has been proposed (Donovan and Scott, 1980).

In contrast, *Botryococcus* remains were found in the Permian Autun shale and are presumed to be the major source of organic matter, except for one sample. No *Botryococcus* remains were found in this sample and the oil produced by its retorting was nearly devoid of the straight-chain alkanes and 1-alkenes which are prominent in oils from *Botryococcus-derived* shales. Evidently, some as yet unidentified algae contributed to the organic matter in this stratum. Biodegradation cannot be ruled out but seems unlikely due to the lack of prominent iso- and ante-iso-alkanes. Straight-chain alkanes and 1-alkenes were also prominent in gas chromatograms of the retorted oils from the Devonian shale. However, in this case, a pronounced hump, which usually indicates polycyclic derivatives, was also prominent. Both extracts and oil from the Devonian shale were found to be rich in steranes and tricyclic compounds. Diterpenoids and triterpenoids have been suggested as precursors for the dicyclic and the tricyclic compounds found in many oil shales. Rock-Eval pyrolysis results indicate that these shales have high hydrogen indices; the kerogens are all type I or type II, with one of the Devonian samples being clearly type I.

Lamosite is pale, grayish-brown, and dark gray to black oil shale in which the chief organic constituent is lamalginite derived from lacustrine planktonic algae. Other minor components include vitrinite, inertinite, telalginite, and bitumen. The Green River oil shale deposits in western United States and a number of the tertiary lacustrine deposits in eastern Queensland, Australia, are lamosites. Other major lacustrine oil shale deposits include the Triassic shales of the Stanleyville Basin in Zaire and the Albert shales of New Brunswick, Canada (Mississippian).

Torbanite, named after Torbane Hill in Scotland, is a black oil shale whose organic matter is composed mainly of telalginite found in freshwater to brackish water lakes. The deposits are commonly small, but can be extremely high grade.

1.2.3 Marine Oil Shale

Marine oil shales (marine-bottom-deposited shales) are composed of lipid-rich organic matter derived from marine algae, acritarchs (unicellular organisms of questionable origin), and marine dinoflagellates.

Marine oil shales are usually associated with one of two settings (Figure 1.1). The anoxic silled basin shown (Figure 1.1a) can occur in the shallow water of a continental shelf. High phytoplankton growth rates near the surface will give a high deposition rate. The sill shields the trough from the circulation of oxygen-laden water. Under these conditions, the decomposition of organic sedimentary matter will rapidly deplete oxygen within the confines of the basin, thereby providing

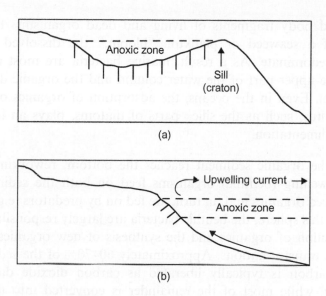

Fig. 1.1. Schematic of an anoxic-silled basin (a) lake shore and (b) sea shore (Scouten, 1990).

the strongly anoxic (reducing, low-Eh) environment that is needed for efficient preservation.

The anoxic zone in an upwelling area (Figure 1.1b) arises from circulation of an open-ocean current over a cold, oxygen-depleted bottom layer. Mixing of a nutrient-rich current, such as the Gulf Stream, into the carbon dioxide-rich and light-rich eutrophic zone gives an environment capable of sustaining very high rates of organic production. Such environments occur today along the west coasts of Africa and the Americas, where good fishing is found along with the potential for organic-rich sediments (Debyser and Deroo, 1969).

Information about the nature of organic matter in marine environments has resulted from studies of recent deposits and the contemporary oceans (Bader et al., 1960; Bordovskiy, 1965). Only a small part of primary production in the oceans reaches the bottom. Of an estimated annual production of 9×10^{19} tons of dry matter, it has been estimated that about 2% reaches the floor in the shallows and only about 0.02% in the open sea. The major part of marine primary production is consumed by predators, and most of the remainder by microbes. The principal marine microbial scavengers are bacteria that live free in the water or are attached to organic particles. In ocean water, organics occur in solution, in colloidal suspension, and as particulate matter comprising bodies and body fragments of living and dead organisms. Except in regions of a seaweed or plankton "bloom," the dissolved organics usually predominate. As a result, marine bacteria are most abundant only in the upper part of the water column and the organic detritus at the bottom. Even in the oceans, the adsorption of organics onto inorganic detritus, such as the silica parts of diatoms, plays an important part in sedimentation.

After the organic sediment reaches the bottom, reworking begins. Bottom-dwelling (benthic) organisms feed on both the sediment and the dissolved organics and, in turn, are fed on by predators (e.g., crustaceans). In this sphere, the benthic bacteria are largely responsible for the decomposition of organics and the synthesis of new organics through enzymatic transformations. Approximately 60–70% of the sedimentary organic carbon is typically liberated as carbon dioxide during this reworking, while most of the remainder is converted into new compounds, resulting in an extremely complex mixture.

The various organic compound classes in oil shales include carbohydrates, lignins, humates and humic acids, lipid-derived waxes and the saturated and polyene acids in algal lipids, which can serve as precursors of these waxes, and biological pigments and their derivatives (e.g., carotenoids and porphyrins). Only the latter three were judged to have sufficient inertness to be major contributors to oil shale kerogen (Cane, 1976).

Proteins, carbohydrates, and humates occur in marine sediments. Protein-derived materials include both original and altered proteins and their decomposition products (amines, amino acids, and amino-complexes). Carbohydrates are rapidly hydrolyzed and are generally not important in oil shales. Humates can be important, even in marine shales, when deposited near the shores, although obviously humic material has been found in marine sediments that were deposited far from land. It is possible that some humic material may be derived from proteins or carbohydrates, perhaps when these materials are adsorbed on inorganic particulates (e.g., clays and volcanic ash) in a moderately oxidizing environment.

Lipids are produced by phytoplanktons and are also synthesized from carbohydrates by microbial activity (Bordovskiy, 1965). With respect to oil shales, the polyene fatty acids are especially interesting.

It is well known that adverse conditions can lead to very high lipid production by algae. At low temperatures and with limited oxygen, for example, *Chiarella* may produce lipids up to >75% of their body weight. Much of these lipids are unsaturated. As expected, due to decarboxylation and some other chemical reactions such as polymerization, polyene acids in lipid fats disappear on heating, while saturated acids remained unaltered. The role of unsaturated lipid acids in the products left by *Botryococcus braunii* blooms in Coorong of southern Australia involves polymerization of the unsaturated lipid residue to produce *coorongite*, a resilient, insoluble material that resembles kerogen in many respects. Iso- and ante-iso-fatty acids have been found in some oil shales, but these are probably secondary products. In fact, bacteria are active in transforming *n*-alkanoic acids into the branched iso- and ante-iso-acids. Other bacterially induced transformations include the hydrogenation of oleic and linoleic acids, the decarboxylation and polymerization of alkanoic acids, as well as lipid hydrolysis.

Carotenoid pigments have been found in many oil shales and in petroleum and coal. Studies of the carotenoids isolated from DSDP (Deep Sea Drilling Project) cores from the Quaternary sediments in the Cariaco Trench show that the chemistry of these materials is largely reductive and traceable over 50 000–350 000 years (Watts and Maxwell, 1977; Watts et al., 1977). This work gives useful insight into the diagenetic transformations of carotenoids, which lead to the observance of partially and perhydrogenated carotenoids in marine oil shale.

The black marine shales formed in shallow seas have been extensively studied as they occur in many places. These shales were deposited on broad, nearly flat sea bottoms and therefore usually occur in thin deposits (10–50 m thick), which may extend over thousands of square miles. The Irati shale (Permian) in Brazil extends over more than 1000 miles from north to south (Costa Neto, 1983). The Jurassic marine shales of western Europe, the Silurian shales of North Africa, and the Cambrian shales of northern Siberia and northern Europe are other examples of this kind of marine oil shale (Tissot and Welte, 1978).

Marinite is a gray to dark gray to black oil shale of marine origin, in which the chief organic components are lamalginite and bituminite derived chiefly from marine phytoplankton. Marinite may also contain small amounts of bitumen, telalginite, and vitrinite. Marinites are deposited typically in an epeiric sea (a sea extending inland from a continental margin) such as on broad shallow marine shelves or inland seas; where wave action is restricted and currents are minimal. The Devonian–Mississippian oil shales of the eastern United States are typical marinites. Such deposits are generally widespread, covering hundreds to thousands of square kilometers, but they are relatively thin, often less than 300 feet.

Tasmanite, named from oil shale deposits in Tasmania, is a brown to black oil shale. The organic matter consists of telalginite derived chiefly from unicellular algae of marine origin and lesser amounts of vitrinite, lamalginite, and inertinite.

Kukersite, which takes its name from Kukruse Manor near the town of Kohtla-Järve, Estonia, is a light brown, marine oil shale. Its principal organic component is telalginite derived from green algae. Kukersite is the main type of oil shale in Estonia and western Russia (Chapter 2).

The organic matter of *kukersite* is considered to be entirely of marine origin and consists almost entirely of accumulations of discrete bodies, telalginite derived from a colonial microorganism termed *Gloeocapsomorpha prisca*. Compared with other rocks containing telalginite, kukersites have low atomic H/C (1.48) and high atomic O/C (0.14) ratios and generally plot as type II kerogen on the van Krevelen diagram (Cook and Sherwood, 1991).

1.3 COMPOSITION AND PROPERTIES

Kerogen is speculated to have originated during the formation of petroleum from sedimentary organic matter. Although this may be true, the follow-on deduction that kerogen is a precursor to petroleum has not been conclusively proven and the theory that the first phase in the transformation of organic matter to petroleum is mere speculation (Speight, 2007).

It is a fact the term *oil shale* describes an organic-rich rock from which little carbonaceous material can be removed by extraction (with common petroleum-based solvents) but which produces variable quantities of distillate (*shale oil*) when raised to temperatures greater than 350°C (660°F). Thus, oil shale is assessed by the ability of the mineral to produce shale oil in terms of gallons per ton by means of a test method (Fischer assay) in which the oil shale is heated to 500°C (930°F).

1.3.1 General Properties

Oil shale is typically a fine-grained sedimentary rock containing relatively large amounts of organic matter (*kerogen*) from which significant amounts of shale oil and combustible gas can be extracted by thermal deposition with ensuing distillation from the reaction zone. However, oil shale does not contain any oil—this must be produced by a process in which the kerogen is thermally decomposed (cracked) to produce the liquid product (shale oil). Thus, any estimate of shale oil reserves can only be speculative, based on estimates from applying the Fischer assay test method to (often) nonrepresentative samples taken from an oil shale deposit. The assay data (in terms of oil yield in gallons per ton) must not to be taken as *proven reserves*.

Kerogen that has not thermally matured beyond the diagenesis (low-temperature) stage occurs typically due to the relatively shallow depth of burial. The Green River oil shale of Colorado has matured to the stage that heterocyclic constituents have formed and predominate, with up to 10% normal paraffins and isoparaffins that boil in the range that includes natural naphtha and gasoline constituents. The relatively high hydrogen/carbon ratio (1.6) is a significant factor in terms of yielding high-quality fuels. However, the relatively high nitrogen content (1–3% w/w) is a major issue in terms of producing stable fuels (petroleum typically contains less than 0.5% nitrogen), as well as producing environmentally detrimental nitrogen oxides during combustion.

In the United States, there are two principal oil shale types: the shale from the Green River Formation in Colorado, Utah, and Wyoming, and the Devonian-Mississippian black shale of the East and Midwest (Table 1.3) (Baughman, 1978). The Green River shale is considerably richer, occurs in thicker seams, and has received the most attention for synthetic fuel production.

The mineral matter (shale) consists of fine-grained silicate and carbonate minerals. The ratio of kerogen-to-shale for commercial grades of oil shale is typically in the range 0.75:5 to 1.5:5—as a comparison, for coal, the organic matter-to-mineral matter ratio in coal is usually greater than 4.75:5 (Speight, 2013).

The common property of these two types of oil shale is the presence of the ill-defined kerogen. The chemical composition of the kerogen has

Table 1.3. Composition (% w/w) of the Organic Matter in the Mahogany Zone and New Albany Shale		
Component (% w/w)	Green River, Mahogany Zone	New Albany
Carbon	80.5	82.0
Hydrogen	10.3	7.4
Nitrogen	2.4	2.3
Sulfur	1.0	2.0
Oxygen	5.8	6.3
Total	100.0	100.0
H/C atomic ratio	1.54	1.08
Source: Baughman (1978).		

been the subject of many studies (Scouten, 1990) but whether or not the data are indicative of the true nature of the kerogen is extremely speculative. Based on solubility/insolubility in various solvents (Koel et al., 2001), it is, however, a reasonable premise (remembering that regional and local variations in the flora that were the precursors to kerogen) led to differences in kerogen composition and properties. Kerogen from different shale samples will differ in composition and properties—similar to the varying in quality, composition, and properties of petroleum from different reservoirs (Speight, 2007).

The organic matter is derived from the varied types of marine and lacustrine algae, with some debris of land plants, and its composition is largely dependent on the depositional environment and sediment sources. Bacterial processes were probably important during the deposition and early diagenesis of most oil shale deposits—these processes could produce significant quantities of biogenic methane, carbon dioxide, hydrogen sulfide, and ammonia. These gases in turn could react with dissolved ions in the sediment waters to form authigenic minerals (minerals generated where they were found or observed) such as calcite ($CaCO_3$), dolomite ($CaCO_3 \cdot MgCO_3$), pyrite (FeS_2), and even rare authigenic minerals such as buddingtonite (ammonium feldspar—$NH_4 \cdot Al \cdot Si_3O_8 \cdot 0.5H_2O$).

The organic matter in oil shale is composed chiefly of carbon, hydrogen, and oxygen with lesser amounts of sulfur and nitrogen. Because of its high molecular weight (best estimates are in the order of several thousands) and molecular complexity, oil shale kerogen is almost totally insoluble in petroleum-based and conventional organic solvents (such as carbon disulfide) (Durand, 1980; Hunt, 1996; Scouten, 1990; Speight, 2007; Tissot and Welte, 1978). A portion of the organic matter in oil shale is soluble and is (incorrectly and confusingly) termed *bitumen*. The bitumen, which is soluble, is dispersed throughout the kerogen network, although even in finely crushed shale much of it may be inaccessible to the solvent. As a result, only a small fraction of the hydrocarbonaceous material in oil shale can be removed by conventional solvent-extraction techniques.

Briefly, the term *bitumen* is more appropriate when applied to the organic content of tar sand (oil sand) deposits, although the name also applies to road asphalt in European countries and other countries (Speight, 2007, 2008). Using this name in reference to the soluble portion of the organic constituents of oil shale is more for convenience than scientific correctness.

Small amounts of bitumen that are soluble in organic solvents are present in some oil shales. Because of its insolubility, the organic matter must be retorted at temperatures in the order of 500°C (930°F) to decompose it into shale oil and gas. After thermal decomposition of the organic matter, some carbon (in the form of a carbonaceous deposit) remains with the shale residue after retorting but can be burned to obtain additional energy.

The organic matter of *kukersite* is considered to be entirely of marine origin and consists almost entirely of accumulations of discrete bodies, telalginite derived from a colonial microorganism termed *G. prisca*. Compared to other rocks containing telalginite, kukersite has a low atomic hydrogen-to-carbon ratio (H/C = (1.48) and high atomic oxygen-to-carbon ratio (O/C = 0.14) and generally falls into the type II kerogen on the van Krevelen diagram (Figure 1.2) (Cook and Sherwood, 1991).

Major components of this kerogen are phenolic moieties with linear alkyl side chains. In spite of the predominance of phenolic moieties, kukersite appears as a highly aliphatic kerogen due to the presence of associated long, linear alkyl chains (Derenne et al., 1989). The formation of kukersite kerogen is believed to have occurred through the selective preservation pathway, and the phenolic moieties correspond to important basic structures of the resistant macromolecular material (Derenne et al., 1994).

Different extraction methods yield bitumen from kukersite on the order of 1 to 3% w/w.

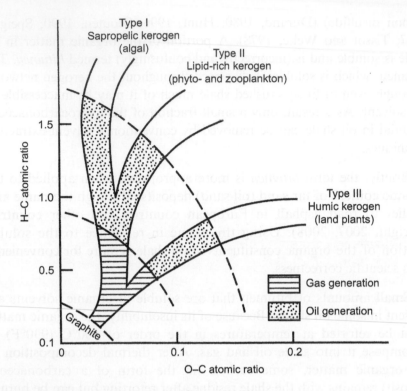

Fig. 1.2. The Van Krevelen diagram showing the different types of kerogen.

The yield of oil and gas under slow retorting conditions is not the same as under Fischer assay. Gas compositions reported for slow, modest pressure retorting indicate that the energy content of the gas could be as much as 70% greater than achieved by Fischer assay (Burnham and Singleton, 1983). This increase has at least three sources of uncertainty: (1) possible leaks in the gas collection system at the slowest heating rate at elevated pressure, (2) difficulty in recovering light hydrocarbons dissolved in the oil at elevated pressure, and (3) the likelihood that oil cracking at higher geological pressures in the liquid phase is less than in the self-purging reactor, which requires volatilization for expulsion. Nevertheless, it is likely that the gas yields will be higher for methane due to oil coking reactions, which was the main reason for the 70% increase, so it is likely that slow retorting would generate gases with good heat content (Burnham, 2003).

Finally, the gross heating value of oil shales on a dry-weight basis ranges from about 500 to 4000 kcal/kg of rock. The high-grade kukersite oil shale of Estonia, which fuels several electric power plants, has a heating value of approximately 2000 to 2200 kcal/kg. By comparison, the heating value of lignite ranges from 3500 to 4600 kcal/kg on a dry-mineral-free basis (Speight, 2013).

1.3.1.1 Mineral Components

Oil shale has often been termed as (incorrectly and for various illogical reasons) high-mineral coal. Nothing could be further from the truth than this misleading terminology. Coal and oil shale are fraught with considerable differences (Speight, 2008, 2013) and such terminology should be frowned upon.

Furthermore, the precursors of the organic matter in oil shale and coal also differ. Much of the organic matter in oil shale is of algal origin, but may also include remains of the vascular land plants that more commonly compose much of the organic matter in coal (Dyni, 2003, 2006; Scouten, 1990; Speight, 2013). In addition, the lack of recognizable biological structures in oil shale that would help identify the precursor organisms makes it difficult to identify the origin of the organic matter.

In terms of mineral and elemental content, oil shale differs from coal in several distinct ways. Oil shale typically contains much larger amounts of inert mineral matter (60–90%) than coal, which has been defined as containing less than 40% mineral matter (Speight, 2013). The organic matter of oil shale, which is the source of liquid and gaseous hydrocarbons, typically has a higher hydrogen and lower oxygen content than that of lignite or bituminous coal.

The mineral component of some oil shale deposits is composed of carbonates including calcite ($CaCO_3$), dolomite ($CaCO_3 \cdot MgCO_3$), siderite ($FeCO_3$), nahcolite ($NaHCO_3$), dawsonite [$NaAl(OH)_2CO_3$], with lesser amounts of aluminosilicates—such as alum [$KAl(SO_4)_2 \cdot 12H_2O$]— and sulfur, ammonium sulfate, vanadium, zinc, copper, and uranium, which add by-product value (Beard et al., 1974). For other deposits, the reverse is true—silicates including quartz (SiO_2), feldspar [$xAl(Al \cdot Si)_3O_8$, where x can be sodium (Na) or calcium (Ca) or potassium (K)], and clay minerals are dominant and carbonates are a minor component. Many oil shale deposits contain small, but ubiquitous, amounts of sulfides including pyrite (FeS_2) and marcasite (FeS_2, but physically and crystallographically distinct from pyrite), indicating that the sediments probably accumulated in dysaerobic (a depositional environment with 0.1–1.0 ml of dissolved oxygen per liter of water) to anoxic waters that prevented the destruction of the organic matter by burrowing organisms and oxidation.

Green River oil shale contains abundant carbonate minerals including dolomite, nahcolite, and dawsonite. The latter two minerals have potential by-product value for their soda ash and alumina content respectively. The oil shale deposits of the eastern United States are low in carbonate content but contain notable quantities of metals, including uranium, vanadium, molybdenum, and others, which could add significant by-product value to these deposits.

There is a potential for low emissions due to the inherent presence of carbonate minerals. Calcium carbonate present in oil shale ash binds sulfur dioxide, and it is not necessary to add limestone for desulfurization:

$$CaCO_3 \rightarrow CaO + CO_2$$

$$2CaO + SO_2 + O_2 \rightarrow CaSO_4$$

Illite (a layered aluminosilicate [K,H$_3$O)(Al,Mg,Fe)$_2$(Si,Al)$_4$O$_{10}$(OH)$_2$, (H$_2$O)] is ever-present in Green River oil shale—it is generally associated with other clay minerals but also frequently occurs as the only clay mineral found in the oil shale (Tank, 1972). Smectite (a group of clay minerals that includes montmorillonite, which tends to swell when exposed to water) is present in all three members of the Green River Formation, but its presence frequently shows an inverse relationship to both analcime (a white, gray, or colorless tectosilicate mineral, which consists of hydrated sodium aluminum silicate, NaAlSi$_2$O$_6 \cdot$ H$_2$O) and loughlinite [a silicate of magnesium, Na$_2$Mg$_3$Si$_6$O$_{16} \cdot$ 8(H$_2$O)]. Chlorite (a group of mostly monoclinic but also triclinic or orthorhombic micaceous phyllosilicate minerals) occurs only in the silty and sandy beds of the Tipton Shale Member. The distribution of random mixed-layer structures and amorphous material is irregular. Several independent lines of evidence favor an in situ origin for many of the clay minerals. Apparently, the geochemical conditions favoring the accumulation of the oil shale also favored the in situ generation of illite.

Finally, precious metals and uranium are contained in good amounts in oil shale of the eastern United States. It may not be viable to recover these mineral resources in the near future, since a commercially favorable recovery process has not yet been developed. However, there are many patents on recovery of alumina from dawsonite-bearing beds [NaAl $(CO_3)(OH)_2$] by leaching, precipitation, and calcination.

1.3.1.2 Thermal Decomposition

Compared to coal, oil shale kerogen is relatively hydrogen-rich and can, therefore, be subjected to thermal conversion leading to higher yields of distillable oil and gas. This is in keeping with volatile products from fossil fuels being related to the hydrogen content of the fossil fuel (Scouten, 1990; Speight, 2007, 2008, 2013).

High-yield oil shale sustains combustion, hence the older Native American name, *the rock that burns*, but in the absence of air (oxygen), three carbonaceous end products result when oil shale is thermally decomposed. Distillable oil is produced as are noncombustible gases, and a carbonaceous (high-carbon) deposit remains on the rock (on the surface or in the pores) as char, a coke-like residue. The relative proportions of oil, gas, and char vary with the pyrolysis temperature and to some extent with the organic content of the raw shale. All three products are contaminated with nonhydrocarbon compounds, and the amounts of the contaminants also vary with the pyrolysis temperature (Bozak and Garcia, 1976; Scouten, 1990).

At temperatures in the order of 500–520°C (930–970°F), oil shale produces shale oil, while the mineral matter of the oil shale is not decomposed. The yield and quality of the products depend on a number of factors, whose impact has been identified and quantified for some of the deposits, notably the US Green River deposits and the Estonian Deposits (Brendow, 2003, 2009; Miknis, 1990). A major factor is that oil shale ranges widely in organic content and oil yield. Commercial grades of oil shale, as determined by the yield of shale oil, range from about 25 to 50 g/t of rock (typically using the Fischer assay method).

The correlation of the shale oil yield with the chemical and physical properties of the oil shale or kerogen has been based on different kinds of measurements, ranging from simple, qualitative tests that can be performed in the field to more complicated measurements in the laboratory.

One simple aspect of the thermal decomposition of oil shale kerogen is the relationships of the organic hydrogen and nitrogen contents and the Fischer assay oil yields. Stoichiometry suggests that kerogen with a higher organic hydrogen-to-carbon atomic ratio can yield more oil per weight of carbon than can kerogen that is relatively hydrogen-poor (Scouten, 1990). However, the hydrogen-to-carbon atomic ratio is not the only important factor. South African kerogen with an atomic hydrogen-to-carbon ratio of 1.35 has a lower oil yield than Brazilian kerogen with an atomic hydrogen-to-carbon ratio of 1.57. In general, the oil shale containing kerogen that can be converted efficiently to oil contains relatively low levels of nitrogen (Scouten, 1990).

During retorting, kerogen decomposes into three organic fractions: (1) shale oil, (2) gas, and (3) carbonaceous residue. Oil shale decomposition begins at relatively low retort temperatures (300°C, 572°F) but proceeds more rapidly and more completely at higher temperatures (Scouten, 1990). The highest rate of kerogen decomposition occurs at retort temperatures of 480–520°C (895–970°F). In general, the shale oil yield decreases, the gas yield increases, and the aromaticity of the oil increases with increasing decomposition temperature (Dinneen, 1976; Scouten, 1990). Furthermore, variation of product distribution with time in the reaction zone can cause a change in product distribution (Figure 1.3) (Hubbard and Robinson, 1950).

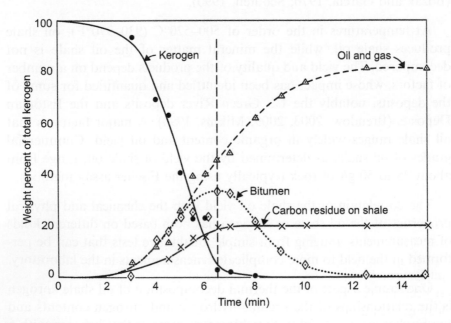

Fig. 1.3. Variation of product yield with time (Hubbard and Robinson, 1950).

However, there is an upper limit on optimal retorting temperature, as the mineral content of the shale may decompose if the temperature is too high. For example, the predominant mineral component of Estonian kukersite shales is calcium carbonate, a compound that dissociates at high temperatures (600–750°C and 1112–1382°F for dolomite; 600–900°C and 1112–1652°F for calcite). Thus, carbon dioxide must be anticipated as a product of oil shale decomposition process, which will dilute the off-gases (adding to emission issues) produced from the retorting process. The gases and vapors leaving the retort are cooled to condense the reaction products, including oils and water.

Kinetically, the active devolatilization of oil shale begins at about 350–400°C (660–750°F), with the peak rate of oil evolution at about 425°C (800°F) and with devolatilization being essentially complete in the range of 470–500°C (890–930°F) (Hubbard and Robinson, 1950; Shih and Sohn, 1980). At temperatures close to 500°C (930°F), the mineral matter, consisting mainly of calcium/magnesium and calcium carbonates, begins to decompose yielding carbon dioxide as the principal product. The properties of crude shale oil are dependent on the retorting temperature, but more importantly on the temperature–time history because of the secondary reactions accompanying the evolution of the liquid and gaseous products. The produced shale oil is dark brown, odoriferous, and tending to waxy oil.

Kinetic studies (Scouten, 1990) indicate that at temperatures below 500°C (930°F), the kerogen (organic matter) decomposes into an extractable product (*bitumen*) with subsequent decomposition into oil, gas, and carbon residue. The actual kinetic picture is influenced by the longer time required to heat the organic material that is dispersed throughout the mineral matrix and the increased resistance to the outward diffusion of the products by the matrix that does not decompose. From the practical standpoint of oil shale retorting, the rate of oil production is the important aspect of kerogen decomposition.

The processes for producing oil from oil shale involve heating (retorting) the shale to convert the organic kerogen to a raw shale oil (Burnham and McConaghy, 2006; Janka and Dennison, 1979; Rattien and Eaton, 1976). Conversion of kerogen to oil without heat has not yet been proven commercially. Although there are schemes for accomplishing such a task, in spite of the claims made about them, they have not moved into the viable commercial or even demonstration stage.

However, since there are issues to consider when using the Fischer assay to determine the potential yields of oil from shale, there are other issues to consider relating to the rate of heating (Allix et al., 2011; Dyni et al., 1989). The reactions that convert kerogen to oil and gas are understood generally, but not in precise molecular detail and can only be represented by simple equations. The amount and composition of generated hydrocarbons depend on the heating conditions, the rate of temperature increase, the duration of exposure to heat, and the composition of gases present as the kerogen breaks down.

Generally, surface-based retorts heat the shale rapidly. The time scale for retorting is directly related to the particle size of the shale, which is why the rock is crushed before being heated in surface retorts. Pyrolysis of particles on the millimeter scale can be accomplished in minutes at 500°C (930°F), while pyrolysis of particles of sizes in tens of centimeters takes much longer.

In situ processes heat the shale more slowly. However, slow heating has advantages—the quality of the oil increases substantially. Coking and cracking reactions in the subsurface tend to leave the higher molecular weight (higher boiling), less desirable components in the ground. As a result, compared with surface processing, an in situ process can produce lighter lower boiling products with fewer contaminants.

From the standpoint of shale oil as a substitute for petroleum products, its composition is of great importance. Oils of paraffin types are similar to paraffin petroleum. However, the composition of the kukersite shale oil of Estonia is more complicated and very specific—it contains abundant oxygen compounds, particularly phenols, which can be extracted from oil. The oil cannot serve directly as raw oil for high-quality engine fuel, but it is easily used as a heating fuel. It has some specific properties, such as lower viscosity, lower pour point, and relatively low sulfur content, which make it suitable for other uses such as marine fuel.

Contrary to other oil shales, obtaining high oil yields of distillable oil from kukersite needs specific conditions of processing. This can be explained by the fact that on thermal processing of kukersite, its elevated moisture percentage and the predominance of calcium carbonate in its mineral part result in high values of specific heat consumption (Yefimov and Purre, 1993). Also, shale is rich in organic matter and must be able to withstand the higher temperatures of thermobitumen formation and coking at a relatively high speed to avoid caking and secondary pyrolysis of oil.

One of the characteristics of kukersite that causes considerable difficulties in its commercial scale processing is its conversion to a bitumen-like product on slow heating—the transition to the plastic state takes place within the temperature range 350–400°C (660–750°F). The maximum yield of thermobitumen is produced at 390–395°C (735–745°F) and it constitutes up to 55–57% w/w of the organic products. At these temperatures, carbon content of solid residue (remaining after extraction with mixture of ethanol–benzene) at a minimum. However, as the heating continues up to 510–520°C (950–970°F), the carbon content of the residue increases two to threefold. As a result, most of the carbonaceous residue in semicoke is of secondary origin, formed by the pyrolysis of unstable components like oxygen-containing compounds (Yefimov and Purre, 1993).

The thermal characterization of Australian oil shale was undertaken by the separation of the unique components of oil shale, the kerogen (organic component) and the clay minerals (inorganic components), using chemical and physical techniques (Berkovich et al., 2000). The heat capacity and the enthalpy changes for kerogen and clay minerals were measured using nonisothermal modulated differential scanning calorimetry from 25°C to 500°C (77–930°F). Heat capacity data was obtained over a temperature range spanning several hundred degrees in a single experiment. Heat capacity was also estimated by incorporating thermogravimetric data during regions where thermal reactions involving mass loss occurred. Enthalpy data for dehydration and pyrolysis of kerogen were also determined (Berkovich et al., 2000; Scouten, 1990).

The mineralogy of Green River Formation is radically changed when the raw oil shale is subjected to extreme processing temperatures (Essington et al., 1987; Milton, 1977; Smith et al., 1978). Mineral reactions from high temperature oil shale retorting can be summarized by two general steps: (1) decomposition of raw minerals and (2) crystallization from the melt. Complete decomposition of carbonate minerals and silicate minerals forms a pyro-metamorphic melt containing the principal ions such as Ca^{2+}, Na^+, Mg^{2+}, Fe^{2+}, Fe^{3+}, K^+, Si^{4+}, Al^{3+}, and O^{2+} (Mason, 2006; Park et al., 1979). Trace elements are abundant in the Green River Formation and are undoubtedly present in the melt, but their low abundance is believed to make their contribution to the crystallization of new minerals negligible, although some partitioning has been recorded (Shendrikar and Faudel, 1978).

Silicate mineral products of high temperature oil shale processing fall into several general types: olivine group, melilite group, ortho- and clin-opyroxenes, amphibole, feldspar group, quartz, and clay minerals. Amorphous silica (glass) is also a common product in oil shale that is processed at high temperatures and then cooled rapidly. Although variation within the mineral groups can be in part due to minor differences in the composition of the raw oil shale, the final mineral suite appears to vary very little when examining material from different processes and localities (Mason, 2006). However, some oil shale deposits contain minerals and metals that add by-products such as dawsonite [NaAl (OH)$_2$CO$_3$], sulfur, ammonium sulfate, vanadium, zinc, copper, and uranium.

1.3.1.3 Oil Shale Grade

The grade of oil shale has been determined by different methods with the results being expressed in various units (Dyni 2003, 2006; Scouten, 1990). For example, the heating value is useful for determining the quality of an oil shale that is burned directly in a power plant to produce electricity. Although the heating value of a given oil shale is a useful and fundamental property of the rock, it does not provide information on the amounts of shale oil or combustible gas that would be yielded by retorting (destructive distillation).

Alternatively, the grade of oil shale can be determined by measuring the yield of distillable oil produced from a shale sample in a laboratory retort (Scouten, 1990). This is perhaps the most common type of analysis that has been, and still is, used to evaluate an oil shale resource—however, the end result of the evaluation depends on the source of the sample and whether or not it is representative of the deposit.

The method commonly used in the United States is the *modified Fischer assay* test method (ASTM D3904). Some laboratories have further modified the Fischer assay method to better evaluate different types of oil shale and different methods of oil shale processing. The standard Fischer assay test method (ASTM D3904, now withdrawn but still used in many laboratories) consists of heating a 100-g sample crushed to 8-mesh (2.38-mm) screen in a small aluminum retort at 500°C (930°F) at a rate of 12°C (21.6°F) per minute and held at that temperature for 40 min. The distilled vapors of oil, gas, and water are passed through a condenser cooled with ice water and then into a graduated centrifuge tube. The oil and water are separated by centrifuging. The quantities reported are the weight percentages of shale oil (and its specific gravity), water, shale residue, and (by difference) gas plus losses.

The Fischer assay method does not measure the total energy content of an oil shale sample because the gases, which include methane, ethane, propane, butane, hydrogen, hydrogen sulfide, and carbon dioxide, can have significant energy content, but are not individually specified (Allix et al., 2011). Also, some retort methods, especially those that heat at a different rate or for different times, or that crush the rock more finely, may produce more oil than that produced by the Fischer assay method. Therefore, the method can only be used as a reference point and, at best, the data from the Fischer assay test method can only be employed to approximate the energy potential of an oil shale deposit.

Other retorting methods, such as the Tosco II process, are known to yield in excess of 100% of that reported by Fischer assay. In fact, some methods of retorting can increase oil yields of some oil shales by as much as three to four times that obtained by the Fischer assay method (Dyni, 2003, 2006; Scouten, 1990).

Another method for characterizing the organic richness of oil shale is a pyrolysis test developed by the Institut Français du Pétrole for analyzing source rocks (Allix et al., 2011). The Rock-Eval test heats a 50–100-mg sample through several temperature stages to determine the amounts of hydrocarbon and carbon dioxide generated. The results can be interpreted for kerogen type and potential for oil and gas generation. The method is faster than the Fischer assay and requires less sample material (Kalkreuth and Macauley, 1987).

1.3.1.4 Porosity

Porosity (void fraction) is a measure of the void spaces in a material such as a reservoir rock and is a fraction of the volume of void space over the total volume, and is expressed as a fractional number between 0 and 1 or as a percentage between 0 and 100.

The porosity of a porous material can be measured in a number of different ways, depending on what specific pores are looked at and how the void volumes are measured. They include (1) interparticle porosity,

(2) intra-particle porosity, (3) internal porosity, (4) porosity by liquid penetration, (5) porosity by saturation, (6) porosity by liquid absorption, (7) superficial porosity, (8) total open porosity, (9) bed porosity—the bed void fraction, and (10) packing porosity.

The porosity of the mineral matrix of oil shale cannot be determined by the methods used to determine the porosity of petroleum reservoir rocks because the organic matter in the shale exists as a solid and is essentially insoluble. However, inorganic particles contain a micropore structure, about 2.36–2.66% v/v, and although the mineral particles have an appreciable surface area, 4.24–4.73 m^2/g for oil shale capable of producing 29–75 g/t in the Fischer assay, the measurement of porosity may be limited to the characteristics of the external surface rather than to the actual pore structure.

Except for the two low-yield oil shale samples, naturally occurring porosities in the raw oil shales are almost negligible and they do not afford access to gases (Table 1.4). Porosity may exist to some degree in the oil shale formation where fractures, faults, or other structural defects have occurred. It is also believed that a good proportion of pores are either blind or inaccessible. Cracking and fractures or other

Table 1.4. Porosity and Permeability of Raw and Treated Oil Shale					
	Porosity			Permeability	
Fischer Assay	Raw	Heated to 815°C	Plane	Raw	Heated to 815°C
1.0[1]	9.0[2]	11.9	A[3]		0.36[4]
			B		0.56
6.5	5.5	12.5	A		0.21
			B		0.65
13.5	0.5	16.4	A		4.53
			B		8.02
20.0	<0.03	25.0	A		
			B		
40.0	<0.03	50.0	A		
			B		

Source: Chilingarian and Yen (1978).
[1] *Fischer assay in gallons per ton.*
[2] *Numbers in percentages of the initial bulk volume. Porosity was taken as an isotropic property, that is, property that is independent of measurement direction.*
[3] *Plane A is perpendicular to the bedding plane; plane B is parallel to the bedding plane.*
[4] *Units in millidarcy.*

structural defects often create new pores and also break up some of the blind pores—closed or blind pores are normally not accessible by mercury porosimetry even at high pressures. Due to the severity of mercury poisoning, the instrument based on pressurized mercury penetration through pores is no longer used.

In the process of the production of shale oil from oil shale, both the chemical and physical properties of oil shale play important roles. The low porosity, low permeability, and high mechanical strength of the oil shale rock matrix make the extraction process less efficient by making the mass transport of reactants and products much harder as well as reducing the process efficiency (Eseme et al., 2007; Scouten, 1990).

Furthermore, the changes in properties as a function of temperature and pressure present implications of the evolution of these properties for in situ exploitation and basin modeling. While the mechanical properties at room temperature are well known, the existing data suggest a positive correlation between oil shale grade (organic matter content) and Poisson ratio, whereas tensile and compressive strength and modulus of elasticity show negative correlations. These properties are strongly affected by temperature—an increase in temperature results in loss of strength and decrease in Young's modulus (Scouten, 1990). Strength follows a logarithmic decrease with increasing temperature, depending on grade. Creep is much enhanced by elevated temperature. Extrapolation of laboratory data to nature suggests that tensile fracturing may occur more easily during petroleum generation, and creep is more prominent in oil shales than in other rocks at this depth in the crust (Eseme et al., 2007).

1.3.1.5 Permeability
Permeability is the ability, or measurement of a rock's ability, to transmit fluids, typically measured in darcies or millidarcies. Permeability is part of the proportionality constant in Darcy's law, which relates the flow rate of the fluid and the fluid viscosity to a pressure gradient applied to the porous media.

The permeability of raw oil shale is essentially zero because the pores are filled with a nondisplaceable organic material. In general, oil shale constitutes a highly impervious system. Thus, one of the major challenges for any in situ retorting project is in the creation of a suitable degree of permeability in the formation. This is why an appropriate rubbelization technique is essential to the success of an in situ pyrolysis project.

Of practical interest is the dependency of porosity or permeability on temperature and organic content. Upon heating to 510°C (950°F), an obvious increase in oil shale porosity is noticed. These porosities, which vary from 3 to 6% v/v of the initial bulk oil shale volume, represent essentially the volumes occupied by the organic matter before the retorting treatment. Therefore, the oil shale porosity increases as the pyrolysis reaction proceeds.

In an oil shale that produces a low yield of oil by the Fischer assay method (lean oil shale), structural breakdown of the cores is insignificant and the porosities are those of intact porous structures. However, in the high-yield Fischer assay oil shale, that is, rich oil shale, this is not the case because structural breakdown and mechanical disintegration due to retorting treatment become extensive and the mineral matrices no longer remain intact. Thermal decomposition of the mineral carbonates, such as magnesium carbonate ($MgCO_3$) and calcium carbonate ($CaCO_3$), actively occurring around 380–900°C (715–1650°F) also results in an increase in porosity.

The increase in porosity from low-yield to high-yield Fischer assay oil shale varies from 2.82 to 50% (Table 1.4). These increased porosities constitute essentially the combined spaces represented by the loss of the organic matter and the decomposition of the mineral carbonates. Cracking of particles also occurs, due to the devolatilization of organic matter that increases the internal vapor pressure of large nonpermeable pores to an extent that the mechanical strength of the particle can no longer contain. Liberation of carbon dioxide from mineral carbonate decomposition also contributes to the pressure buildup in the oil shale pores.

1.3.1.6 Compressive Strength
Raw oil shale has high compressive strengths both perpendicular and parallel to the bedding plane. After heating, the inorganic matrices of low-yield Fischer assay oil shale retain high compressive strength in both perpendicular and parallel planes. This indicates that a high degree of inorganic cementation exists between the mineral particles comprising each lamina and between adjacent laminae. With an increase in organic matter of oil shale, the compressive strength of the respective

organic-free mineral matrices decreases, and it becomes very low in those rich oil shale.

1.3.1.7 Thermal Conductivity

Measurements of thermal conductivity of oil shale show that blocks of oil shale are anisotropic about the bedding plane and thermal conductivity as a function of temperature, oil shale assay and direction of heat flow, parallel to the bedding plane (parallel to the earth's surface for a flat oil shale bed), was slightly higher than the thermal conductivity perpendicular to the bedding plane. As layers of sedimentary material were deposited to form the oil shale bed over geological time, the resulting strata have a higher resistance to heat flow perpendicular to the strata than parallel to the strata (Table 1.5).

The thermal conductivity of retorted and burnt shale is lower than those of the raw shales from which they are obtained (Table 1.5). This is attributable to the fact that the mineral matter is a better conductor of heat than the organic matter, and, on the other hand, the organic matter is still a far better conductor than the voids created by its removal. While the first of the above hypotheses is well justified when one takes into account the contribution of the lattice conductivity to the overall value, the effect of the amorphous carbon formed from the decomposition of the organic matter could also be of importance in explaining the difference in thermal conductivity values for retorted shales and the corresponding burnt samples. The role of voids in

Table 1.5. Comparison between Thermal Conductivity Values for Green River Oil Shale			
Temperature Range (°C)	Fischer Assay (gallons per ton)	Plane	Thermal Conductivity (J/m/s/°C)
38–593	7.2–47.9	—	0.69–1.56 (raw shales)
			0.26–1.38 (retorted shales)
			0.16–1.21 (burnt shales)
25–420	7.7–57.5	A	0.92–1.92
		Average	1.00–1.82 (burnt shales)
38–205	10.3–45.3	A	0.30–0.47
		B	0.22–0.28
20–380	5.5–62.3	A	1.00–1.42 (raw shales)
		B	0.25–1.75 (raw shales)
A—parallel to the bedding plane; B—perpendicular to the bedding plane; Average—average of both directions.			

determining the magnitude of the effective thermal conductivity is likely to be significant only for samples with high organic contents.

The thermal conductivity of oil shale is, in general, only weakly dependent on the temperature. However, extreme caution needs to be exercised in the interpretation of results at temperatures close to the decomposition temperature of the shale organic matter. This is due to the fact that the kerogen decomposition reaction (or pyrolysis reaction) is endothermic in nature, and as such the temperature transients can be confounded between the true rate of heat conduction and the rate of heat of reaction.

REFERENCES

Allix, P., Burnham, A., Fowler, T., Herron, M., Kleinberg, R., Symington, B., 2011. Coaxing Oil from Shale. Oilfield Rev. 22 (4), 6.

Allred, V.D. (Ed.), 1982. Oil Shale Processing Technology. Center for Professional Advancement, East Brunswick, NJ.

Andrews, A., 2006. Oil Shale: History, Incentives, and Policy. Specialist, Industrial Engineering and Infrastructure Policy Resources, Science, and Industry Division. Congressional Research Service, the Library of Congress, Washington, DC.

Bader, R.G., Hood, D.H., Smith, J.B., 1960. Recovery of dissolved organic matter in sea-water and organic sorption by particulate material. Geochim. Cosmochim. Acta 19, 236–243.

Bartis, J.T., LaTourrette, T., Dixon, L., 2005. Oil Shale Development in the United States: Prospects and Policy Issues. Prepared for the National Energy Technology of the United States Department of Energy. Rand Corporation, Santa Monica, CA.

Baughman, G.L., 1978. Synthetic Fuels Data Handbook, second ed. Cameron Engineers, Inc., Denver, CO.

Beard, T.M., Tait, D.B., Smith, J.W., 1974. Nahcolite and Dawsonite Resources in the Green River Formation, Piceance Creek Basin, Colorado. Guidebook to the Energy Resources of the Piceance Creek Basin, 25th Field Conference, Rocky Mountain Association of Geologists, Denver, CO, pp. 101–109.

Berkovich, A.J., John, H., Levy, J.H., Schmidt, S.J., Young, B.R., 2000. Heat capacities and enthalpies for some Australian oil shales from non-isothermal modulated DSC. Thermochim. Acta 357–358, 41–45.

Bordovskiy, O.K., 1965. Accumulation and transformation of organic substances in marine sediments. Mar. Geol. 3, 3–114.

Bozak, R.E., Garcia Jr., M., 1976. Chemistry in the oil shales. J. Chem. Educ. 53 (3), 154–155.

Brendow, K., 2003. Global oil shale issues and perspectives. Oil Shale 20 (1), 81–92.

Brendow, K., 2009. Oil shale—a local asset under global constraint. Oil Shale 26 (3), 357–372.

Burnham, A.K., 2003. Slow Radio-Frequency Processing of Large Oil Shale Volumes to Produce Petroleum-like Shale Oil. Report No. UCRL-ID-155045. Lawrence Livermore National Laboratory, US Department of Energy, Livermore, CA.

Burnham, A.K., McConaghy, J.R., 2006. Comparison of the acceptability of various oil shale processes. In: Proceedings of AICHE 2006 Spring National Meeting, Orlando, FL, March 23–27, 2006.

Burnham, A.K., Singleton, M.F., 1983. High-pressure pyrolysis of Green River oil shale. In: Miknis, F.P. (Ed.), Chemistry and Geochemistry of Oil Shale. Symposium Series No. 230, Washington, DC, pp. 335–351.

Cane, R.F., 1976. The origin and formation of oil shale. In: Yen, T.F., Chilingarian, G.V. (Eds.), Oil Shale. Elsevier, Amsterdam, Netherlands.

Chilingarian, G.V., Yen, T.F., 1978. Bitumens, Asphalts, and Tar Sands. Elsevier, Amsterdam, Netherlands (Chapter 1).

Cook, A.C., Sherwood, N.R., 1991. Classification of oil shales, coals and other organic-rich rocks. Org. Geochem. 17 (2), 211–222.

Culbertson, W.C., Pitman, J.K., 1973. Oil Shale in United States Mineral Resources, Paper No. 820. United States Geological Survey, Washington, DC.

Debyser, J., Deroo, G., 1969. Observations on the genesis of petroleum. Rev. Institut Français du Pétrole. 24 (1), 21–48.

Derenne, S., Largeau, C., Casadevall, E., Sinninghie Damste, J.S., Tegelaar, E.W., deLeeuw, J.W., 1989. Characterization of Estonian kukersite by spectroscopy and pyrolysis: evidence for abundant alkyl phenolic moieties in an Ordovician, marine, type II/I kerogen. Org. Geochem. 16 (4–6), 873–888.

Derenne, S., Largeau, C., Landais, P., Rochdi, A., 1994. Spectroscopic features of *Gloeocapsomorpha prisca* colonies and of intersyitial matrix in kukersite as revealed by transmission micro-FT-IR: location of phenolic moieties. Fuel 73 (4), 626–628.

Dinneen, G.U., 1976. Retorting technology of oil shale. In: Yen, T.F., Chilingar, G.V. (Eds.), Oil Shale. Elsevier Science Publishing Company, Amsterdam, Netherlands, pp. 181–198.

Donovan, R.N., Scott, J., 1980. Lacustrine cycles, fish ecology and stratigraphic zonation in the Middle Devonian of Caithness. J. Geol. 16, 35–50.

Durand, B., 1980. Kerogen: Insoluble Organic Matter from Sedimentary Rocks. Editions Technip, Paris, France.

Dyni, J.R., 2003. Geology and resources of some world oil-shale deposits. Oil Shale 20 (3), 193–252.

Dyni, J.R., 2006. Geology and Resources of Some World Oil Shale Deposits. Report of Investigations 2005-5295. United States Geological Survey, Reston, VA.

Dyni, J.R., Anders, D.E., Rex, R.C., 1989. Comparison of hydroretorting, Fischer assay, and Rock-Eval analyses of some world oil shales. In: Proceedings of 1989 Eastern Oil Shale Symposium. Institute for Mining and Minerals Research, University of Kentucky, Lexington, KY, pp. 270–286.

Eseme, E., Urai, J.L., Krooss, B.M., Littke, R., 2007. Review of mechanical properties of oil shales: implications for exploitation and basin modelling. Oil Shale 24 (2), 159–174.

Essington, M.E., Spackman, L.K., Harbour, J.D., Hartman, K.D., 1987. Physical and Chemical Characteristics of Retorted and Combusted Western Reference Oil Shale. Report No. DOE/MC/11076-2453. United States Department of Energy, Washington, DC.

Hubbard, A.B., Robinson, W.E., 1950. A Thermal Decomposition Study of Colorado Oil Shale. Report of Investigations No. 4744. United States Bureau of Mines, Washington, DC.

Hunt, J.M., 1996. Petroleum Geochemistry and Geology, second ed. W.H. Freeman, San Francisco, CA.

Hutton, A.C., 1987. Petrographic classification of oil shales. Int. J. Coal Geol. 8, 203–231.

Hutton, A.C., 1991. Classification, organic petrography and geochemistry of oil shale. In: Proceedings of 1990 Eastern Oil Shale Symposium. Institute for Mining and Minerals Research, University of Kentucky, Lexington, KY, pp. 163–172.

Janka, J.C., Dennison, J.M., 1979. Devonian Oil Shale in Symposium Papers: Synthetic Fuels from Oil Shale, Atlanta, GA, December 3–6, pp. 21–116.

Kalkreuth, W.D., Macauley, G., 1987. Organic petrology and geochemical (Rock-Eval) studies on oil shales and coals from the Pictou and Antigonish areas, Nova Scotia, Canada. Bull. Can. Pet. Geol. 35, 263–295.

Koel, M., Ljovin, S., Hollis, K., Rubin, J., 2001. Using neoteric solvents in oil shale studies. Pure Appl. Chem. 73 (1), 153–159.

Lee, S., 1996. Alternative Fuels. Taylor & Francis Publishers, Washington, DC.

Mason, G.M., 2006. Fractional differentiation of silicate minerals during oil shale processing: a tool for prediction of retort temperatures. In: Proceedings of 26th Oil Shale Symposium, Colorado School of Mines, Golden, CO, October 16–19.

Miknis, F.P., 1990. Conversion characteristics of selected foreign and domestic oil shales. In: Proceedings of 23rd Oil Shale Symposium. Colorado School of Mines, Golden, CO, pp. 100–109.

Milton, C., 1977. Mineralogy of the Green River formation. The Mineralogical Record, 8, 368–379.

Ots, A., 2007. Estonian oil shale properties and utilization in power plants. Energetika 53 (2), 8–18.

Park, W.C., Linderamanis, A.E., Robb, G.A., 1979. Mineral changes during oil shale retorting. In Situ 3 (4), 353–381.

Rattien, S., Eaton, D., 1976. Oil shale: the prospects and problems of an emerging energy industry. In: Hollander, J.M., Simmons, M.K. (Eds.), Annual Review of Energy, vol. 1, pp. 183–212.

Scouten, C., 1990. Part IV: Oil Shale. Chapters 25–31. In: Speight, J.G. (Ed.), Fuel Science and Technology Handbook. Marcel Dekker Inc., New York.

Shendrikar, A.D., Faudel, G.B., 1978. Distribution of trace metals during oil shale retorting. Environ. Sci. Technol. 12 (3), 332–334.

Shih, S.M., Sohn, H.Y., 1980. Non-isothermal determination of the intrinsic kinetics of oil generation from oil shale. Ind. Eng. Chem. Process Des. Dev. 19, 420–426.

Smith, J.W., Robb, W.A., Young, N.B., 1978. High temperature mineral reactions of oil shale minerals and their benefit to oil shale processing in place. In: Proceedings of 11th Oil Shale Symposium, Colorado School of Miners, Golden, CO, pp. 100–112.

Speight, J.G., 2007. The Chemistry and Technology of Petroleum, fourth ed. CRC-Taylor and Francis Group, Boca Raton, FL.

Speight, J.G., 2008. Synthetic Fuels Handbook: Properties, Processes, and Performance. McGraw-Hill, New York.

Speight, J.G., 2013. The Chemistry and Technology of Coal, third ed. CRC-Taylor and Francis Group, Boca Raton, FL.

Tank, R.W., 1972. Clay minerals of the Green River formation (Eocene) of Wyoming. Clay Miner. 9, 297.

Tissot, B., Welte, D.H., 1978. Petroleum Formation and Occurrence. Springer-Verlag, New York.

US DOE, 2004a. Strategic Significance of America's Oil Shale Reserves, I. Assessment of Strategic Issues, March. http://www.fe.doe.gov/programs/reserves/publications

US DOE, 2004b. Strategic Significance of America's Oil Shale Reserves, II. Oil Shale Resources, Technology, and Economics; March. http://www.fe.doe.gov/programs/reserves/publications

US DOE, 2004c. America's Oil Shale: A Roadmap for Federal Decision Making; USDOE Office of US Naval Petroleum and Oil Shale Reserves. http://www.fe.doe.gov/programs/reserves/publications

Wang, D-M., Xu, Y-M., He, D-M., Guan, J., Zhang, O-M., 2009. Investigation of mineral composition of oil shale. Asia-Pac. J. Chem. Eng. 4, 691–697.

Watts, R.L., Maxwell, J.R., 1977. Carotenoid diagenesis in a marine sediment. Geochim. Cosmochim. Acta 41, 493–497.

Watts, R.L., Miller, R.C., Kjosen, H., 1977. The potential of carotenoids as environmental indicators. In: Campos, R., Goni, J. (Eds.), Advances in Organic Geochemistry, 1975, Enadimsa, Madrid, Spain, pp. 391–413.

Yefimov, V., Purre, T., 1993. Characteristics of kukersite oil shale, some regularities and features of its retorting. Oil Shale 10 (4), 313–319.

CHAPTER 2

Oil Shale Resources

2.1 INTRODUCTION

Oil shale deposits are found on all continents and such deposits contain a solid hydrocarbonaceous material (*kerogen*) (Chapter 1) that can be converted to crude shale oil by thermal decomposition (Chapters 4 and 5).

In regard to local (United States) deposits of oil shale, the deposits that occur in the western United States have a different chemical composition and geological history from the deposits in the eastern United States. In addition, oil shale deposits found elsewhere—in countries such as Australia, Brazil, Estonia, People's Republic of China, Russia, Scotland, and Spain—also offer chemical and geological differences to the deposits found in the United States. Nevertheless, these countries have been the sites of small-scale oil shale industries in the past, and various levels of government and private industries have recognized the potential of oil shale as a source of liquid fuel and still continue attempts to establish an oil shale industry or are giving this idea a serious consideration (Brendow, 2003, 2009).

In this regard, oil shale is quite different from petroleum, which is more concentrated in certain regions of the world. Depending upon the data source and the year of reporting, the statistical values may vary. In fact, before progressing to *resources* and *reserves* of various countries, it is necessary to first understand that *in-place resources* and *proved reserves* (*proven reserves*) have markedly different meanings (Speight, 2007, 2008, 2011, 2013).

The former (*in-place resources*) relates to *potential reserves*, while the latter (*proved reserves* or *proven reserves*) indicates the existence of fossil fuel resources that can be exploited. For example, a deposit of oil shale with economic potential is typically one that is at or near the surface and hence can be developed by open-pit or conventional underground mining or by in situ methods (Dyni, 2003, 2006; Scouten, 1990). Although some Colorado oil shale reaches the surface at places, such as the Colony deposit, many deposits typically start at approximately 1000 feet beneath the surface and extend downward for as much as another 2000 feet.

Within the Colorado oil shale column are rocks that vary considerably in kerogen content, with some portions of the section having a higher kerogen content (richer oil shale) and other portions of the deposit having a lower kerogen content (leaner oil shale). The entire column has been estimated to be able to produce on the order of one million barrels of oil equivalent per acre (BOE/acre) over its productive life—but that is only an estimate and is based on many factors, some of which are out of the control of the developer owing to chemical and geological factors and may be subject to politics more than to an accurate assessment of the resources (Speight, 2011).

Equating oil shale with other types of hydrocarbon fossil fuel and hydrocarbon-producing fossil fuels often indicates (to some observers) that comparable means of exploiting oil shale can be used. However, oil shale is not directly comparable to either crude oil or coal or even tar sand bitumen, though it may appear to share some characteristics with both (Speight, 2008). Each ton of oil shale—a carbonate rock, generally mudstone or siltstone (Chapter 1)—contains significant quantities of a solid organic sediment (conveniently referred to as *kerogen*) and a trace of extractable organic material (conveniently but incorrectly referred to as *bitumen*) and gas. Some oil shale is rich in carbonate (marl or marlstone—a calcium carbonate—rich or lime-rich mud or mudstone, which contains variable amounts of clay and silt), whereas other deposits are relatively rich in clay minerals (Dyni, 2003, 2006; Scouten, 1990). All of which serves to make oil shale resource estimation (especially the potential for producing shale oil) much more difficult.

Indeed, the reliability of resource data, as indicated in the foregoing paragraphs, can range from excellent to poor. Data for some deposits that have been explored extensively by core drilling, such as the Green River oil shale in Colorado, the kukersite deposit in Estonia, and some of the Tertiary deposits in eastern Queensland, Australia, are especially good in comparison to others. Many other resource data are open to question, speculation, and (at best) inspired guesswork.

Thus, a word of caution is advised. The numbers presented from geological studies of the deposits in the various countries are a mixed bag.

Evaluation of world oil shale resources is especially difficult—even more difficult than the estimation of petroleum or coal resources (Speight, 2011, 2013)—because of the wide variety of analytical units that are reported. Furthermore, some resources are presented as reserves of in situ shale oil—at best only a potential and highly speculative number—while others are presented as percent by weight of in-place organic matter, again an estimated number but still better than the former number since none of the formations contain shale oil.

2.2 TOTAL RESOURCES

The potential resources of oil shale in the world are enormous, but a precise evaluation, like petroleum resources, is difficult because of the numerous ways by which the resources are assessed (Speight, 2011). Although many oil shale deposits have been explored only to a minor extent, some deposits have been fairly well delineated by drilling and analyses (Dyni, 2003, 2006; Scouten, 1990; Johnson et al., 2004). These include the Green River oil shale in the western United States, the Tertiary deposits in Queensland, Australia, the deposits in Sweden and Estonia, the El-Lajjun deposit in Jordan, and some of the deposits in Brazil, France, Germany, and Russia. The remaining deposits are poorly delineated, and further competent investigation and analysis are needed to adequately determine their resource potential.

The largest known deposit is the Green River Formation in the western United States, which is a total estimated in-place resource that *may* have the potential to produce approximately 3 trillion barrels (3×10^{12} bbls) of shale oil. In Colorado alone, the total in-place resource is supposed to be on the order of 1.5 trillion barrels (1.5×10^{12} bbls) of oil. The Devonian black shales of the eastern United States are estimated to have the potential to produce 189 billion barrels (189×10^9 bbls) of oil.

The total world in-place resource of oil shale has the tongue-in-cheek potential to produce 4.8 trillion barrels (4.8×10^{12} bbls) of shale oil. The actual data could be much higher (or even much lower) since the oil shale resources of some countries are not reported and many known deposits have not been fully investigated. On the other hand, several deposits, such as those of the Heath and Phosphoria Formations and portions of the Swedish alum oil shale, have been degraded by geothermal heating.

Therefore, the resources reported for such deposits are probably too high and somewhat misleading. The amount of shale oil that can be recovered from an oil shale deposit (recoverable resources) depends upon many factors, the least of which are (1) the character of the deposit and (2) the method used for assessment.

To determine the former (i.e., character of the deposit) requires careful drilling and geochemical investigation. The method used for the latter (i.e., assessment of the potential to produce shale oil) is always open to question. Gravimetric, volumetric, and heating values have all been used to determine the oil shale grade, which is usually expressed in gallons per ton (or liters per tonne) if the grade of oil shale is given in volumetric measure (gallons per ton, tonne); the specific gravity (density) of the oil must be known to convert gallons to a weight percent estimate (Dyni, 2003, 2006; Scouten, 1990; Speight, 2008). A deposit worthy of commercial investigation might be expected to produce at least 10 gallons per ton of shale oil.

The obvious need is for new and improved methods for the economic recovery of shale oil and any added-value products, but the deciding factor for oil shale development has always been governed by the price of petroleum. The fluctuating price of petroleum (usually fluctuating to the high side of the price range) and the geopolitics of recent decades have prompted governments around the world to reexamine national energy supplies and to consider national security issues (Speight, 2011). All seem to have reached the same conclusions: energy security can be accomplished only by developing indigenous natural resources (such as oil shale).

At present, oil shale is commercially exploited in several countries: Brazil, China, Estonia, and Australia. Brazil has a long history of oil shale development, and it is known that oil shale in Brazil has been exploited since the late nineteenth century. In China, 80 new retorts (Fushun retorts) are used to produce shale oil (Qian et al., 2003). There are also claims that China has the fourth largest oil shale deposits in the world after the United States, Brazil, and Russia.

2.3 OCCURRENCE AND HISTORY BY COUNTRY

Oil shale represents a large and mostly untapped source of hydrocarbon fuel. Like tar sands (*oil sands* in Canada), it is an unconventional or alternate fuel source and it does not contain oil. Oil is produced by thermal decomposition of kerogen, which is intimately bound within the mineral matrix of the shale and, as such, is not readily extractable.

Many estimates have been published for oil shale reserves (in fact resources), but the rank of countries vary with time and authors, except that the United States is always number one with more than 60% of the world reserves. Brazil is most frequently at number two. The United States has vast known oil shale resources that could translate into as much as 2.2 trillion barrels (2.2×10^{12} bbls) of *oil-in-place*. In fact, the largest known oil shale deposits in the world are in the Green River Formation, which covers portions of Colorado, Utah, and Wyoming. Estimates of the oil resource within the Green River Formation vary from 1.5 to 1.8 trillion barrels (1.5 to 1.8×10^{12} bbls) (Dyni, 2006; Scouten, 1990). However, not all resources in place are recoverable but, for policy planning purposes, it is enough to know that any amount in this range is very high. For example, the half-way point in the estimate (800–900 billion barrels [800 to 900×10^9 bbls]) is more than triple the proven oil reserves of Saudi Arabia. With present demand for petroleum products in the United States running at approximately 17–20 million barrels per day, oil shale (by only meeting a quarter of that demand) would last for more than 400 years.

Oil shale occurs in nearly 100 major deposits in 27 countries worldwide (Bauert, 1994; Culbertson and Pitman, 1973; Culbertson et al., 1980; Duncan and Swanson, 1965). It requires generally shallower (<3000 feet) zones than the deeper and warmer geological zones required to form oil. On a worldwide basis, the oil shale resource base is believed to contain about 2.6 trillion barrels (2.6×10^{12} bbls) of shale oil, of which the vast majority (eastern plus western shale)—approximately 2.2 trillion barrels (2.2×10^{12} bbls)—is located within the United States.

2.3.1 Australia

Australia has significant potential unconventional oil resources contained in oil shale deposits in several basins (Figure 2.1), which range from small and noneconomic deposits to some large enough for commercial development. The demonstrated oil shale resources of Australia total 58 billion tons (58×10^9 tons), of which approximately 24 billion barrels (24×10^9 bbls) of shale oil is recoverable (Australian Government, 2010; Cook and Sherwood, 1989; Crisp et al., 1987).

Production from oil shale deposits in southeastern Australia began in the 1860s, but came to an end in 1952 when government funding ceased.

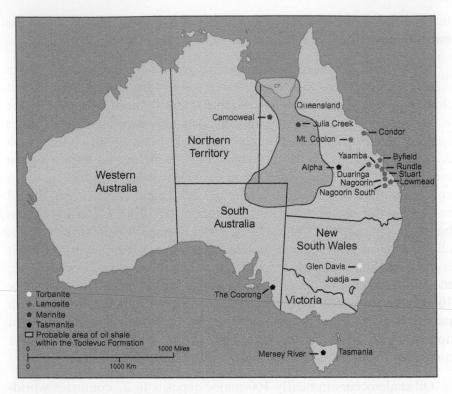

Fig. 2.1. Australian oil shale deposits.

Between 1865 and 1952, some 4 million tonnes (4.4×10^6 U.S. tons) of oil shale were processed. Much of the early production of oil shale in Australia was from the torbanite deposits of New South Wales. During the early years of mining, torbanite was used for gas enrichment in Australia and overseas, but paraffin, kerosene, and wood preserving and lubricating oils were also produced. Of 30 deposits in New South Wales, 16 were commercially exploited (Crisp et al., 1987).

Two small deposits of torbanite have been investigated in Queensland. These include the small but high-grade Alpha deposit, which constitutes a potential in situ resource of 19 million U.S. barrels (19×10^6 U.S. bbls) (Noon, 1984, p. 4) and a smaller deposit at Carnarvon Creek. Several companies attempted to develop the marine Tasmanite deposits of Permian age in Tasmania during the early 1900s. Between 1910 and 1932, a total of 1100 m^3 (approximately 7600 barrels) of shale oil was produced from several intermittent operations. Further developments are unlikely unless new resources are found (Crisp et al., 1987).

Oil shale in the marine Toolebuc Formation of Early Cretaceous age underlies approximately 18 000 square miles (approximately 12 000 000 acres) in parts of the Eromanga and Carpenteria Basins in Queensland and adjacent states. The oil shale zone ranges up to 25 feet in thickness but yields on average only approximately 10 gallons per ton. However, the Toolebuc Formation is estimated to contain the potential for approximately 1.7 trillion barrels (1.7×10^{12} bbls) of in situ shale oil that could be produced by open-pit mining of the shale (Ozimic and Saxby, 1983; Sherwood and Cook, 1983). The atomic hydrogen to carbon (H/C) ratio of the organic matter is about 1.1 $\pm$ 0.2 with high aromaticity (>50%). However, only approximately 25% of the organic matter converts to oil by conventional retorting (Ozimic and Saxby, 1983), which substantially reduces to potential yield of shale oil.

As a result of the increase in the price of crude oil during the oil crisis of 1973 and 1974, exploration for oil shale in Australia was greatly accelerated during the late 1970s and early 1980s. However, by 1986, the prices of crude oil dropped dramatically and interest in the exploitation of oil shale diminished (Crisp et al., 1987).

During the late 1970s and early 1980s, a modern exploration program was undertaken by two Australian companies, Southern Pacific Petroleum NL and Central Pacific Minerals NL (SPP/CPM). The aim was to find high-quality oil shale deposits amenable to open-pit mining operations in areas near infrastructure and deep-water ports. The program was successful in finding a number of silica-based oil shale deposits of commercial significance along the coast of Queensland. Ten deposits clustered in an area north of Brisbane were investigated and found to have an oil shale resource in excess of 20 billion barrels (20×10^9 bbls) (based on a cutoff grade of 50 l/t at 0% moisture), which could support production of more than 1 million barrels per day of shale oil.

Since 1995, Southern Pacific Petroleum NL and Central Pacific Minerals NL (continuing their interest and at one time joined by the Canadian oil sand company Suncor Energy Inc.) have been studying the Stuart Deposit near Gladstone, Queensland, which has an estimated potential to produce 2.6 billion barrels (2.6×10^9 bbls) of oil. From June 2001 until March 2003, 703 000 barrels of oil, 62 860 barrels of light fuel oil, and 88 040 barrels of ultra-low sulfur naphtha were produced from the Gladstone area. After processing, the shale oil could be converted to a low-emission gasoline. Suncor had the role of operator of the Stuart project and in April 2001, Southern Pacific Petroleum and Central Pacific Minerals purchased Suncor's interest.

Having committed itself to ensuring that the Stuart oil shale project had a sustainable development, Southern Pacific Petroleum put various schemes into operation to achieve its stated environmental goals. One in particular launched in 1998 was a reforestation carbon dioxide sink. Approximately, 250 000 trees were planted on deforested lands in Central Queensland. In September 2000, the first carbon trade in Queensland was announced between Southern Pacific Petroleum and the state government and was based on the reforestation trials.

The Stuart project (found to have a total in situ shale oil resource of 2.6 billion barrels and a capacity to produce more than 200 000 b/d) incorporated the Alberta Taciuk Processor retort technology into the project (Schmidt, 2003), which involved a revolving kiln (originally developed for Alberta tar sand projects). The project had three stages: (1) the demonstration plant, which produced a 42° API gravity shale oil containing 0.4% w/w sulfur and 1.0% w/w nitrogen, was constructed between 1997 and 1999 and produced over 500 000 barrels of oil product. The plant was designed to process 6000 tonnes (6600 U.S. tons) per day of run-of-mine (wet shale) and to produce 4500 barrels per day of shale oil products; (2) the process was to be scaled up by a factor of 4 to a commercial-sized module processing 23 500 tonnes (27 600 U.S. tons) per day to produce 15 500 barrels per day of shale oil—it was envisaged that multiple commercial Taciuk processor units would come on stream during 2010–2013, giving the plant the ability of processing up to 380 000 tonnes (41 900 U.S. tons) per day and producing up to 200 000 barrels per day shale for a period in excess of 30 years; and finally (3) the third stage was conceived as processing 125 000 tonnes (138 000 U.S. tons) per day of oil shale to give 65 000 barrels per day of shale oil products, bringing total Stuart production to about 85 000 barrels per day. Overall, shale oil production ran for 87 days of operation and peaked at 3700 barrels per day.

However, the project ended when Southern Pacific Petroleum was placed in receivership in 2003. In February 2004, Queensland Energy Resources Ltd. (QERL) acquired the oil shale assets of Southern Pacific Petroleum and ran final plant trials at the demonstration facility. However, no production ensued and the Environmental Protection

Agency regulated operations until the plant was closed in mid-2004. The facility is now on *care and maintenance* in an operable condition. Queensland Energy Resources continues to assess the possibilities for future commercial operation of the Stuart project. The company (QERL) spent the 2005–2007 period testing indigenous Australian oil shale at a pilot plant in Colorado and demonstrated successfully that, by using the Paraho process, it could operate an oil shale-to-shale oil and liquid products business in Queensland.

In August 2008, the Queensland Government announced that it had issued a 20-year moratorium on the development of the McFarlane oil shale resource by Queensland Energy Resources. This deposit, located some 10 miles south of Proserpine in central Queensland, is considered a strategically important resource, with the potential to supply in excess of 1.6 billion barrels (1.6×10^9 bbls) of oil. Following more than a quarter of a century of extraction of test material, Queensland Energy Resources announced during the third quarter of 2009 that it had reached agreement with the Queensland Government to backfill and rehabilitate the McFarlane box cut.

Also in 2008, it was announced by Queensland Energy Resources that the company had decided against the use of the Taciuk processor because of potential problems in scaling-up the technology to commercial size. Queensland Energy Resources decided in favor of the Paraho II technology to develop oil shale deposits along Queensland's east coast. The company also noted that the oil shale deposits, collectively known as the McFarlane oil shale, have the potential to produce 1.6 billion barrels (1.6×10^9 bbls) of shale oil over the next 40 years.

During 2009, Queensland Energy Resources undertook refurbishment of the site and dismantled the Taciuk retort. In May 2010, the company (QERL) announced the construction of a demonstration plant at Yarwun, north of Gladstone. Using Paraho II technology (Chapter 4), the plant when complete is expected to process 2.5 tonnes (2.8 U.S. tons) of shale per hour and produce between 37 and 40 barrels per day of synthetic crude oil.

In September 2011, Queensland Energy Resources Ltd (QERL) produced its first crude oil from its demonstration Paraho II vertical shaft kiln processing plant at the Stuart deposit near Gladstone, central Queensland. Further news is awaited with eager anticipation.

2.3.2 Brazil

The Brazilian oil shale resource base is one of the largest in the world, and nine deposits of oil shale ranging from Devonian to Tertiary age have been reported in different parts of Brazil (Padula, 1969). Of these, two deposits have received the most interest: (1) the lacustrine oil shale of Tertiary age in the Paraíba Valley in the State of São Paulo, northeast of the city of São Paulo, and (2) the marinite oil shale of the Permian Iratí Formation, which outcrops extensively in southern Brazil, and contains estimated reserves of more than 700 million barrels of oil and 880 billion cubic feet (880×10^9 ft^3) of gas.

The Paraíba Valley deposits contain an estimated reserve of oil shale with the potential to produce 840 million barrels (840×10^6 bbls) of shale oil, and the total resource is estimated to be 2 billion barrels (2×10^9 bbls). The deposit (approximately 145-foot thick) includes several types of oil shale: (1) brown to dark brown fossiliferous laminated paper shale, (2) semi-papery oil shale of the same color, and (3) dark olive, sparsely fossiliferous, low-grade oil shale that fractures semi-conchoidally (Dyni, 2006).

The Iratí oil shale is dark gray, brown, and black, very fine-grained, and laminated. Clay minerals compose 60–70% of the shale and organic matter makes up much of the remainder, with minor amounts of detrital quartz, feldspar, pyrite, and other minerals. Carbonate minerals are sparse. The Iratí oil shale is not notably enriched in metals, unlike marine oil shales such as the Devonian oil shales of eastern United States.

Brazil started production of shale oil in 1881 and is ranked second after the United States for resources (well distributed) and after Estonia for production. In 1935, shale oil was produced at a small plant in Sìo Mateus do Sul in the State of Paran, and in 1950, following government support, a plant capable of producing 10 000 barrels per day shale oil was proposed for Trememb, Sìo Paulo. Brazil developed the world's largest surface oil shale pyrolysis reactor – the Petrosix 35 feet (11 meters) vertical shaft gas combustion retort.

The oil shale resource base is substantial and was first exploited in the late nineteenth century in the State of Bahia (Dyni, 2006). Further demonstration plants were built in the 1970s and 1980s, and shale oil production has continued since that time.

After the development of the Petrosix process for shale oil production, operations were concentrated on the reservoir of Sìo Mateus do Sul and a pilot plant (8-inch internal diameter retort) was brought into operation in 1982. A 6-foot internal diameter retort demonstration plant followed in 1984 and was used for the optimization of the Petrosix technology. A 2200 tons per day, 18-foot (internal diameter) prototype retort (the Irati Profile Plant), originally brought on line in 1972, began operating on a limited commercial scale in 1981 and a further commercial plant—a 35-foot internal diameter retort—was brought into service in December 1991.

Surface facilities at Sao Mateus do Sul, in the state of Parana, are capable of processing 7800 U.S. tons of shale per day to produce fuel oil, naphtha, liquefied petroleum gas, shale gas, sulfur, and asphalt additives. As near as can be determined, the Petrosix retorting process (Petrosix), where the shale undergoes pyrolysis, yields a nominal daily output of approximately 3870 barrels of shale oil along with fuel gas and sulfur.

There are reports that the intention of Petrobras is to maintain the technological expertise and development of its indigenous capacity but without expansion.

2.3.3 Canada

Oil shale occurs throughout Canada and 19 oil shale deposits have been identified. However, the majority of the in-place oil shale resources remain poorly defined—the most explored deposits are those in the provinces of Nova Scotia and New Brunswick (Ball and Macauley, 1988; Hyde, 1984; Kalkreuth and Macauley, 1987; Macauley, 1981).

The Devonian–Lower Mississippian oil shale deposits in the eastern United States extend north into Ontario and Quebec and have a high volume of organic-rich shale (Matthews, 1983; Matthews et al., 1980). A similar setting also occurs in the Ordovician in Canada, but the total volume of organic-rich shale is lower. Flooding of the continental margin also occurred in Arctic Canada in the Devonian and Ordovician, and organic-rich shales of these ages are present in the Northwest Territories. In eastern Canada, a series of rift basins developed in Nova Scotia, New Brunswick, and Newfoundland in the Carboniferous era. In several of these basins, intervals of organic-rich lacustrine shales developed, but the total volume of the oil shale had not been fully defined and the potential of the oil shale to produce shale oil may be limited (Hyde, 1984).

The oil shale deposits range from Ordovician to Cretaceous in age and include deposits of lacustrine and marine origin; as many as 19 deposits have been identified (Davies and Nassichuk, 1988; Macauley, 1981). During the 1980s, a number of the deposits were explored by core drilling (Macauley, 1981, 1984a, b; Macauley et al., 1985; Smith and Naylor, 1990).

Outcrops of Lower Carboniferous lacustrine oil shale (Grinnell Peninsula, Devon Island, Canadian Arctic Archipelago) are as much as 300-foot thick and yield up to 100 gallons per ton of shale. However, for most Canadian deposits, the resources of in situ shale oil remain poorly known (Dyni, 2006).

Of the areas in Nova Scotia known to contain oil shale, development has been attempted at two deposits—Stellarton and Antigonish. Mining took place at Stellarton from 1852 to 1859 and during 1929 and 1930 and at Antigonish in 1865. The Stellarton Basin is estimated to hold some 825 million tonnes (909×10^6 U.S. tons) of oil shale, with a hypothetical in situ oil content of approximately 168 million barrels. The Antigonish Basin has the second largest oil shale resource in Nova Scotia, with an estimated 738 million tonnes (813×10^6 U.S. tons) of shale capable of producing a speculative 76 million barrels of shale oil.

Investigations of the retorting and co-combustion (with coal for power generation) of Albert Mines oil shale (New Brunswick) have been conducted, including some experimental processing in 1988 at the Petrobras plant in Brazil. Interest has been shown in the New Brunswick deposits for the potential they might offer to reduce sulfur emissions by co-combustion of carbonate-rich shale residue with high-sulfur coal in power stations.

In mid-2006, Altius, a Canadian company based in Newfoundland, was awarded a license to explore for oil shale in the Albert Mines prospect. During 2008 and 2009, a drilling program was undertaken within a license area of 240 acres. It is believed that the oil shale resource is likely to be significant, but detailed evaluation is necessary before any reasonable estimates can be made.

Development of the Alberta oilfields and the Athabasca (Alberta) tar sand deposits and Lloydminster (Alberta-Saskatchewan border) has taken precedence (Speight, 2007, 2009), and the need to develop the oil shale deposits has only continued on a low level of effort.

2.3.4 China

The two main resources of oil shale occur in Fushun (Liaoning Province) and Maoming (Guangdong Province) (Baker and Hook, 1979; Shi, 1988).

In the Fushun area (Jijuntun Formation), extensive layers of lacustrine oil shale (49- to 190-foot thick, 15- to 58-m thick) are mined along with coal, both from Eocene lacustrine deposits. The oil yield of the shale ranges from approximately 5–16% w/w of the shale, and the mined shale averages 19–25 gallons per ton shale oil. In the vicinity of the mine, oil shale resources are estimated at 260 million tonnes, of which 235 million tonnes (90%) are considered mineable. The total resource of oil shale at Fushun is estimated at 3.6 billion tonnes (3.6×10^9 U.S. tons) (Dyni, 2006; Fang et al., 2008).

The oil shale in the Jijuntun Formation can be divided into two parts of differing composition: (1) the lower 49 feet (15 m) of light-brown oil shale of low-grade and (2) the upper 330 feet (100 m) of brown to dark-brown, finely laminated oil shale. The oil content of the low-grade oil shale is less than 4.7% w/w and the richer upper grade is greater than 4.7% w/w. However, depending on the exact location of the deposit, the maximum oil content can be as high as 16% w/w. It has been reported that the average oil production is on the order of 78–89 L of oil per tonne of oil shale (assuming a 0.9 specific gravity).

The Maoming oil shale deposit (Youganwo Formation) has total reserves on the order of 5 billion tons (5×10^9 tons) oil shale, of which 860 million tons are in the Jintang mine. The Fischer assay yield of shale oil is 4–12% w/w of the oil shale. The Eocene Maoming oil shale occurs as a laterally uniform stratigraphic section and the sediments included lignite, a vitrinite lens from the overlying claystone, and four intervals from the massive oil shale deposit.

Oil shale retorting was carried out in Fushun, Manchuria, in 1929 with the construction of the first of three plants. After World War II, Refinery No. 1 had 200 retorts, each with a daily throughput of 100–200 tonnes of oil shale. It continued to operate and was joined by the Refinery No. 2 starting up in 1954. In Refinery No. 3, shale oil was hydrotreated to produce light liquid fuels. Shale oil was also open-pit mined in Maoming, Guangdong Province, and 64 retorts were put into operation there in the 1960s.

A second plant began production in 1954 and a third facility began producing shale oil at Maoming in 1963. At the beginning of the 1960s, 266 retorts were operating in Fushun Refineries Nos. 1 and 2. However, by the early 1990s, the availability of much cheaper crude oil led to the Maoming operation and Fushun Refineries No. 1 and 2 being shut down. It has been estimated that production of the Chinese oil shale industry in the mid-1970s was expanded to 55 000–80 000 barrels per day of shale oil. A new facility, the Fushun Oil Shale Retorting Plant, came into operation in 1992 under the management of the Fushun Bureau of Mines. Sixty Fushun-type retorts, each having a capacity of 100 tons of oil shale per day, produced approximately 415 000 barrels of shale oil per year at Fushun (Zhou, 1995).

Between 2004 and 2006, China undertook its first national oil shale evaluation, which confirmed that the resource was both widespread and vast. According to the evaluation, it was estimated that a total oil shale resource of some 720 billion tonnes (795×10^9 U.S. tons) is located across 22 provinces, 47 basins, and 80 deposits. Approximately 70% of the deposits are in eastern and middle China, with the remainder largely in the Qinghai-Tibet area and the west. The in-place shale oil resource has been estimated at some 48 billion tonnes (approximately 354×10^9 barrels) of shale oil.

During 2007, the Fushun Mining Group Co. was operating 180 retorts, each capable of processing 100 tonnes of oil shale per day. The shale ash by-product is utilized to produce building materials. At the beginning of 2010, it was reported that a 6000 t/d Taciuk retort, imported by Fushun and placed in service by the end of 2009, had been delayed. Many other retorts are either operating or being planned in the provinces of Gansu, Guangdong, Hainan, Heilongjiang, and Jilin.

In 2008, Chinese shale oil production totaled approximately 7600 barrels per day, and in 2010, the production of shale oil was reputed to have risen and it has been estimated that production will rise to 10 000 barrels per day. Furthermore, several companies are involved in researching new retorting technologies for processing pulverized or particulate oil shale, with the possibility of constructing a pilot-scale demonstration plant (Liang, 2006; Qian et al., 2003).

Development of the oil shale sector in China has been sustained partly because of the high level of imports of petroleum and petroleum products, necessary to support indigenous demand with the creation of a middle class and the accompanying demand for automobiles. China also prefers to utilize a national resource in the face of high international oil prices. This competes with the goal of seeking higher production from any one of several heavy oil fields, which has served to keep shale oil production levels low.

2.3.5 Egypt

Oil shale was discovered during the 1940s as a result of oil rocks self-igniting while phosphate mining was taking place. The phosphate beds in question lie adjacent to the Red Sea in the Safaga-Quseir area of the Eastern Desert.

Analysis was at first undertaken in the Soviet Union in 1958 and was followed by further research in Berlin in the late 1970s. This latter work concentrated on the phosphate belt in the Eastern Desert, the Nile Valley, and the southern Western Desert. The results showed that the Red Sea area was estimated to have about 4.5 billion barrels of in-place shale oil and that in the Western Desert, the oil shale in Abu Tartour area (which could be mined while mining for phosphates) had the potential to produce up to 1.2 billion barrels (1.2×10^9 bbls) of shale oil.

Although assessment of the oil shale resources continue to establish (or estimate) the potential of the Egyptian resource, the Egyptian government has signed a joint agreement with Jordan, Morocco, Syria, and Turkey, together with regional and international companies, to develop the center with the aims of providing a joint environmental and energy framework, developing common standards for studying and utilizing oil shale resources, and attracting investors to the sector.

While the agreement indicates that the center was to be headquartered in Jordan, the future of the Center and agreement remain unknown at the time of writing as a result of the recent unrest in Egypt and Syria.

2.3.6 Estonia

The Baltic Oil Shale Basin is situated near the north-western boundary of the East European Platform (Figure 2.2) (Baker and Hook, 1979; Lippmaa and Maramäe, 1999, 2000, 2001; Loog et al., 1996).

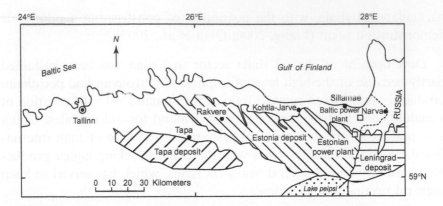

Fig. 2.2. Estonian oil shale deposits (Dyni, 2006).

The Estonia and Tapa deposits are both situated in the west of the Basin, the former being the largest and highest-quality deposit within the Basin. Estonian oil shale resources are currently put at 5 billion tonnes (5.5×10^9 U.S. tons), including 1.5 billion tonnes (1.65×10^9 U.S. tons) of active (mineable) reserves. It is possible that the power production part of the industry will disappear by 2020 and that the resources could last for 30–50 years, but scenarios abound on the replacement of oil shale by alternative resources.

The *kukersite* shale is the most important mineral resource of Estonia. There are two principal deposits in the republic. The first is the *Estonian* located in the northeastern part of the republic. The productive seam thickness diminishes from approximately 10 feet in the northern part of deposit to approximately 6 feet in the southern and western parts. The second principal deposit is the *Tapa* deposit situated southwest of the *Estonian* deposit with seam depth of 200–600 feet below the surface. The seam has maximum thickness of 6–7 feet in the central part of the deposit (Reinsalu, 1998).

Estonian kukersite contains three main components: organic (kerogen), carbonaceous, and terrigenous matter. The last two constitute kukersite mineral matter. The kerogen content (organic matter of kukersite) varies from 10 to 65% w/w. The content of carbon in the organic matter of kukersite is low (76.7% w/w) and the oxygen and carbon mass ratio is 0.13. Estonian kukersite has a high content of hydrogen (9.7% w/w) and a low content of nitrogen (0.3% w/w) in the organic portion of the oil shale. The hydrogen-to-carbon (H/C) atomic ratio is on the order of 1.25, but the sulfur content in the organic matter of kukersite is 1.6% w/w.

Estonian oil shale was first scientifically researched in the eighteenth century (Kattai and Lokk, 1998). In 1838, work was undertaken to establish an opencast pit near the town of Rakvere and an attempt was made to obtain oil by distillation. Although it was concluded that the rock could be used as solid fuel and, after processing, as liquid or gaseous fuel, the *kukersite* (derived from the name of the locality) was not exploited until the fuel shortages created by World War I began to impact.

Permanent mining began in 1918 and has continued until the present day, with capacity (both underground mining and opencast) increasing as demand rose. By 1955, oil shale output had reached 7 million tonnes $(7.7 \times 10^6$ U.S. tons) and was mainly used as a power station/chemical plant fuel and in the production of cement. The opening of the 1400 MW Baltic Thermal Power Station in 1965 was followed, in 1973, by the 1600 MW Estonian Thermal Power Station, which again boosted production and by 1980 (the year of maximum output) the figure had risen to 31.35 million tonnes $(31.35 \times 10^6$ tonnes). However, in 1981, the fourth reactor of the Sosnovy Bor nuclear power station opened in the Leningrad District (Russia), thereby reducing demand for Estonian shale. As a result, production of Estonian oil shale has decreased and remains at a lower level than the production in 1980. The decline lasted until 1995, with nominal annual increases in production since that time.

In December 2009, after a construction period of 2½ years, a new 3000 tonne per day oil shale processing plant was officially opened. Located in Kohtla-Järve, the plant is designed to produce more than 100 000 tonnes of shale oil, 30 million m^3 of high-calorific gas, and 150 GWh of steam. Eesti Energia Technology Industries (operating as Enefit) is currently constructing a 2.26 million tonne per year oil shale plant in Narva. The plant, planned to produce 290 000 t/year of oil is due to start up in 2012. Three additional Enefit 280 units and an upgrader plant are scheduled to be started in 2013.

The Estonian government has taken steps toward privatization of the oil shale industry and is beginning to tackle the air and water pollution problems that nearly a century of oil shale processing has brought. In 1999, 10.7 million tonnes $(11.8 \times 10^6$ U.S. tons) of oil shale were produced. Imports amounted to 1.4 million tonnes (1.5×10^6 U.S. tons), 0.01 million tonnes (0.011×10^6 U.S. tons) were exported, 11.1 million tonnes (12.2×10^6 U.S. tons) were used for electricity and heat generation, and 1.3 million tonnes (1.4×10^6 U.S. tons) were distilled to produce approximately 950 000 barrels of shale oil.

Until recently, only 16% of Estonian shale was used for petroleum and chemical manufacturing. However, because of environmental problems, the goal is to decrease oil shale production.

At the time of writing, Estonia is the only country in the world that operates oil shale-fired power plants to supply most of its electricity to domestic customers and to export power to neighboring countries. In addition to thermal power plants, Estonia also has oil shale thermal processing plants for shale oil production. Power plants and processing factories in Estonia are supplied with oil shale from two underground mines and two opencast mines (Ots, 2007).

Furthermore, Eesti Energia (also called Enefit, outside of Estonia), the company responsible for Estonia's more than 90 years of oil shale mining and 50 years of oil shale surface retort production, has moved into the United States—on March 30, 2011, Enefit American Oil purchased 100% of Oil Shale Exploration Company (OSEC). With the purchase of OSEC, Enefit acquired one of the largest tracts of privately owned oil shale in the United States, totaling more than 30 000 acres and reputed to contain sufficient oil shale to produce approximately 2.1 billion barrels (2.1×10^9 bbls) of shale oil. The company plans to develop a mining, retorting, and upgrading project to produce 50 000 barrels per day of refined shale oil—the first out-of-the-pipe oil is scheduled for 2020.

2.3.7 Ethiopia

The existence of oil shale deposits in Ethiopia has been known since the 1950s. Although surveys have been undertaken in the past, no projects were proceeded with owing to high mining costs and lack of funding.

In 2006, it was reported that the resource, estimated to be 3.89 billion tonnes, in the northern province of Tigray is considered to be suitable for opencast mining. In the Ethiopian Year 2000 (July 2007–June 2008), the Geological Survey of Ethiopia undertook surveys in the Sese Basin, western Ethiopia, to establish the nature and content of the oil shale (and coal) deposits. A certain amount of analysis has been carried out but further research is required.

2.3.8 France

Oil shale was irregularly exploited in France between 1840 and 1957, but its highest (1950) output reached only 0.5 million tonnes (0.6×10^6 U.S. tons) per year of shale. During its 118 year life of the project, the Government imposed taxes and duties on foreign oil, thus preserving the indigenous industry.

In 1978, it was estimated that the in-place shale oil resources amounted to 7 billion barrels (7×10^9 bbls). In mid-2009, Toreador Resources Corporation reported that it had a four-phase plan to exploit the oil shale of the Paris Basin.

2.3.9 Germany

The German oil shale industry was developed in the middle of the nineteenth century, and during the 1930s and 1940s, the development of retorted oil contributed to the depleted fuel supplies during World War II.

In 1965, it was estimated that Germany's in-place shale oil resources amounted to 2 billion barrels (2×10^9 bbls) of shale oil. In recent years, only a minimal quantity of oil shale (0.5 million tonnes per annum, 0.6×10^6 U.S. tons) was produced for use at the Rohrback cement works at Dotternhausen, where it is consumed directly as a fuel for power generation, and the residue is used in the manufacture of cement. At the beginning of 2004, Holcim, a Swiss cement and aggregates company, acquired Rohrbach Zement.

The oil shale from this area is a low energy mineral (with a low oil yield and a high ash content), but by using a complex process, the complete utilization of both the oil shale energy and all its minerals can be accomplished and incorporated into the manufacture of cement. The heat of this process is used simultaneously to produce electricity.

2.3.10 India

Although oil shale, in association with coal and also oil, is known to exist in the far northeastern regions, the extent of the resource and the quality of the oil shale has not yet been determined.

Currently, oil shale, recovered with coal during the mining process, is discarded as a waste product. However, the Indian Directorate General of Hydrocarbons has initiated a project designed to assess the reserve and its development. The project will cover geological mapping, sampling, and analysis in the states of Assam and Arunachal Pradesh. Feasibility and environmental impact assessment studies have also been planned.

2.3.11 Indonesia

Faced with declining reserves of oil and gas, Indonesia has accelerated its research into identifying, and possibly utilizing, indigenous oil shale resources.

The Center for Geo Resources is currently engaged in surveying and preparing an inventory of occurrences. To date, three main prospective oil shale areas have been found, two on the island of Sumatera and one on Sulawesi.

2.3.12 Israel

Marinite deposits of Late Cretaceous age have been identified in Israel (Fainberg and Hetsroni, 1996; Minster, 1994), containing approximately 12 billion tons (12×10^9 tons) of oil shale reserves with an anticipated shale oil yield of 6% w/w of the oil shale. The organic content of the oil shales is relatively low, ranging from 6% to 17% w/w, with an oil yield of 15–17 gallons per ton. Israeli oil shale is generally relatively low in heating value and oil yield and high in sulfur content compared with other major deposits.

Sizeable deposits of oil shale have been discovered in various parts of Israel, with the principal resources located in the north of the Negev desert. Estimates of the theoretical reserves total some 300 billion tonnes (330×10^9 U.S. tons), of which those considered to be open-pit mineable are put at only a few billion tonnes. The largest deposit (Rotem Yamin) has shale beds with a thickness of 35–80 m, yielding 60–71 L of oil per tonne. Generally, Israeli oil shale is relatively low in heating value and shale oil yield and high in moisture, carbonate, and sulfur content compared with other major deposits.

A commercially exploitable bed of phosphate rock, 25- to 50-foot thick, underlies the oil shale in the Mishor Rotem open-pit mine and some of the deposits can be mined by open-pit methods. The largest deposit (Rotem Yamin) has shale beds with a thickness of 100–250 feet, yielding 15–20 gallons of shale oil per ton. A pilot power plant fuelled by oil shale has been technically proven in the Negev region. Annual production of oil shale has averaged approximately 450 000 tonnes (500×10^3 U.S. tons) in recent years.

Following tests in a 0.1 MW pilot plant (1982–1986), a 1 MW demonstration fluidized-bed pilot plant was established in 1989. In operation since 1990, the generated energy is sold to the Israeli Electric Corporation and the low-pressure steam to an industrial complex, and a considerable quantity of the resulting ash is used to make products such as cat litter, which is exported to Europe.

During 2006, A.F.S.K. Hom-Tov, an Israeli company, presented a scheme to the Ministry of National Infrastructures for the manufacture of synthetic oil from oil shale. The method would entail combining bitumen (from the Ashdod refinery, 80 km north of the proposed plant at Mishor Rotem in the Negev Desert) with the shale before processing in a catalytic converter.

Although the Government is encouraging development of the oil shale resource, particularly in situ underground techniques, it is mindful of the environmental concerns. While the country investigates the possibilities of harnessing its large oil shale deposits for producing shale oil, some of the resource is utilized directly for the production of electricity.

2.3.13 Jordan
Jordan is ranked eighth among 37 countries in the world shale oil reserves—more than 65 billion tons (65×10^9 tons) have been recorded over all Jordan, of which 50 billion tons (50×10^9 tons) are located in central Jordan.

Oil shale in northern Jordan was recognized for the first time in the early twentieth century in the Yarmouk region, north of Jordan, near Al-Maqqarin Village. During the First World War the German Army used it when they installed the first project to produce oil from oil shale to operate the Hijazi Railway. Exploration work started after the El Lajjun deposit had been discovered by the German Geological Mission in the 1960s. Intensive exploration activities on oil shale in central Jordan were carried out during the 1980s and resulted in delineating other deposits such as Sultani, Hasa and Jurf Ed Darawish. Continued exploration resulted in the discovery of other deposits such as Attarat Um Ghudran, Wadi Maghar, Siwaqa, Khan El Zabib, and El Thammad (Alali, 2006; Jaber et al., 1997).

In all, there are at least 24 known occurrences of oil shale, which result in Jordan having an extremely large proven and exploitable oil shale resource. Geological surveys indicate that the existing shale reserves cover more than 60% of the country and amount to in excess of 40 billion tonnes. The proven and exploitable reserves of oil shale occur in the central and north-western regions of the country (Alali, 2006; Bsieso, 2003; Hamarneh, 1998; Jaber et al., 1997). The major deposits of commercial-scale interest are located about 60 miles south of Amman (Bsieso, 2003).

Jordanian shale is generally of quite good quality, with relatively low ash and moisture content. Oil yield (5–12% w/w) is favorably comparable with the oil yields from the oil shale of western Colorado (USA); however, Jordanian shale has an exceptionally high sulfur content (up to 9% by weight of the organic content). The reserves are exploitable by opencast mining and are easily accessible (Bsieso, 2003).

The principal mineral component of the oil shale is calcite or more rarely quartz together with kaolinite and apatite and, on occasion, feldspar, muscovite, illite, goethite, and gypsum as secondary components. Dolomite ($CaCO_3.MgCO_3$) occurs in some individual carbonate beds as in the Arbid limestone of El-Lajjun. The main elements of the oil shale, if organic carbon is excluded, are calcium and silicon; minor constituents are sulfur, aluminum, iron, and phosphorous. The concentrations of the remaining components are generally low. The silicon is derived from two sources: clastic sediment input together with titanium, aluminum, and iron and from sedimentary or early diagenetic silicification.

The amount of phosphorous in the shale increases from the top to the bottom of the sequence. Phosphorous content is not favorable for the utilization of the spent shale for the manufacture of cement. However, a certain percentage of the oil shale and the spent shale can be used in cement manufacturing. Molybdenum, chromium, and tungsten are significantly enriched in the bituminous marl in comparison to limestone. Zinc, vanadium, nickel, copper, lanthanum, and cobalt are also enriched, whereas barium is depleted and the content of arsenic and lead are low to moderate. The uranium content is relatively high, but it is clearly associated with phosphorous and not with the bituminous organic matter. The sulfur content ranges from 0.3 to 4.3% w/w (Alali, 2006).

The eventual exploitation of the only substantial fossil fuel resource to produce liquid fuels or electricity, together with chemicals and building materials, would be favored by three factors: (1) the high organic-matter content of Jordanian oil shale, (2) the suitability of the deposits for surface-mining, and (3) their location near potential consumers (i.e., phosphate mines, potash, and cement works).

In May 2010, Enefit (Eesti Energia) signed a concession agreement with the Jordanian Government granting the former the right to utilize part of the Attarat Um Ghudran deposit for 50 years. Located in central Jordan and estimated to contain 25 billion tonnes, the deposit is considered to be the largest in the country. Enefit will undertake further geological research and an environmental impact assessment. After a maximum period of 4 years, a decision will be taken regarding the economic feasibility of the project. If commercial development ensues, it is planned that a 900 MW (maximum) capacity oil shale fired power plant will begin operating in 2016 and a 38 000 barrels per day shale oil plant in 2017.

2.3.14 Kazakhstan

The occurrence of oil shale is widespread and the most important deposits have been identified in western Kazakhstan (the Cis-Urals group of deposits) and eastern Kazakhstan (the Kenderlyk deposit). Further deposits have been discovered in both the southern region (Baikhozha and the lower Ili river basin) and the central region (the Shubarkol deposit).

In excess of 10 deposits have been studied: the Kenderlyk Field has been revealed as the largest (in the region of 4 billion tonnes, 4.4×10^9 U.S. tons) and has undergone the greatest investigation. However, studies on the Cis-Urals group and the Baikhozha deposit have shown that they have important concentrations of rare elements (rhenium and selenium), providing all these deposits with promising prospects for future industrial exploitation.

The shale oil resources have been estimated as having the potential to produce approximately 2.8 billion barrels (2.8×10^9 bbls) of shale oil. Moreover, many of the deposits occur in conjunction with hard and brown coal accumulations which, if simultaneously mined, could increase the profitability of the coal production industry, while helping to establish a shale-processing industry.

At the beginning of the 1960s, successful experimentation was carried out on a sample of Kazakhstan's oil shale in the former Soviet Republic of Estonia. Both domestic gas and shale oil were produced. It was found that the resultant shale oil had a sufficiently low sulfur content for the production of high-quality liquid fuels.

Beginning in early 1998 and lasting until the end of 2001, a team funded by INTAS (an independent, international association formed by the European Community to preserve and promote scientific co-operation with the newly independent states) undertook a project aimed at completely reevaluating the oil shale resources of Kazakhstan. The resultant report testified that Kazakhstan's oil shale resources could sustain the production of various chemical and power-generating fuel products.

In September 2009, it was reported that a high-level bilateral economy, science, and technology cooperation agreement had been signed by Estonia and Kazakhstan. Estonia expressed a willingness to share its expertise in the field of oil shale to help Kazakhstan develop its own resource.

2.3.15 Mongolia
Mongolia possesses large mineral deposits which, owing to the country's political isolation during most of the twentieth century, remain largely undeveloped. Some mining operations were established before 1989 with the help of the Soviet Union and Eastern European countries, but following the breakup of the USSR, Mongolia's move to a free economy and the passage of a Minerals Law in 1997, the potential is being recognized.

Numbered amongst the indigenous minerals are oil shale deposits from the Lower Cretaceous Dsunbayan Group, located in the east of the country. Exploration and investigation of the deposits began as long ago as 1930, but it was only during the 1990s with the help of Japanese organizations that detailed analyses began. Twenty-six deposits were studied and found to be associated with coal measures.

During 2004, the Narantuul Trade Company, the owner of the Eidemt deposit, was investigating the possibilities of developing the field's potential with the aid of international cooperation.

It was reported in the late 2006 that China University of Petroleum had signed a contract to undertake a feasibility study on the Khoot oil shale deposit.

2.3.16 Morocco

Morocco has substantial oil shale reserves, but to date they have not been exploited to any great extent. The total oil shale resource of Morocco is estimated to have the potential to produce some 50 billion barrels $(50 \times 10^9 \text{ bbls})$ of shale oil, an amount which (if proven) would rank the country amongst the world leaders in respect of shale oil (Bouchta, 1984).

Exploitation of oil shale in Morocco occurred as long ago as 1939, when the Tanger deposit was the source of fuel for a pilot plant (88 U.S. tons per day), which operated until 1945. A preliminary estimate of this resource has been put at some 2 billion barrels of oil in place.

During the 1960s, two important deposits were located: (1) Timahdit in the region of the Middle Atlas range of mountains (north central Morocco) and (2) Tarfaya in the south west, along the Atlantic coast. Shale oil potential has been estimated at approximately 16 billion barrels $(16.0 \times 10^9 \text{ bbls})$ for Timahdit and 22.7 billion barrels $(22.7 \times 10^9 \text{ bbls})$ for Tarfaya.

During the early 1980s, Shell and the Moroccan state entity ONAREP conducted research into the exploitation of the oil shale reserves at Tarfaya, and an experimental shale-processing plant was constructed at another major deposit (Timahdit) (Bekri, 1992; Bouchta, 1984). At the beginning of 1986, however, it was decided to postpone shale exploitation at both sites and to undertake a limited program of laboratory and pilot-plant research.

The technical and economic feasibility studies have resulted in Morocco acquiring a large amount of information—a database that can be used for future projects. With the current need to look at developing alternative sources of liquid fuels, government has stated that any pilot plant should be followed by a demonstration phase during which the commercial evaluation of by-products should also be undertaken.

2.3.17 Nigeria

Research has shown that the southeastern region of Nigeria possesses a low-sulfur oil shale deposit. The reserve has been estimated to be of the order of 5.76 billion tonnes (6.3×10^9 U.S. tons) with the potential to produce 1.7 billion barrels (1.7×10^9 bbls) of shale oil.

An oil shale deposit, possibly of high economic value and corresponding to the Turonian Ezeaku shale (lower Nkalagu formation) of the Lower Benue Trough, was found in a $1.5 \times 1.0 \text{ km}^2$ belt in Lokpanta near Okigwe, in Imo State, Nigeria (Ekweozor and Unomah, 1990). The characteristically dark-grey, laminated, and fissile marlstone contains total organic carbon (TOC) in excess of 7% w/w in some locations and total extractable organic matter generally in excess of 10 000 ppm. The kerogen is type I–II (oil-prone), and at the updip rim, it has attained intermediate thermal maturity status. An initial appraisal of the economic potential of the fossil fuel deposit by pyrolysis (modified Fischer assay) indicates an average oil-yield on the order of 40 gallons per ton.

2.3.18 Russia

In excess of 80 oil shale deposits have been identified in Russia. There are oil shale deposits in Leningrad Oblast, across the border from those in Estonia. Annual output is estimated to be about 2 million tonnes (2.1×10^6 U.S. tons), most of which is exported to the Baltic power station in Narva, Estonia. In 1999, Estonia imported 1.4 million tonnes (1.5×10^6 U.S. tons) of Russian shale but is aiming to reduce the amount involved or eliminate the trade entirely. There is another oil shale deposit near Syzran on the river Volga (Kashirskii, 1996; Russell, 1990).

Russia has been mining its reserves on a small-scale basis since the 1930s, with the oil shale being used to fuel two power plants, but the operation has been abandoned owing to environmental pollution. However, most activity has centered on the Baltic Basin where the kukersite oil shale has been exploited for many years.

The exploitation of Volga Basin shale, which has a higher content of sulfur and ash, began in the 1930s. Although the use of such shale as a power-station fuel has been abandoned owing to environmental pollution, a small processing plant may still be operating at Syzran in 1995, with a throughput of less than 50 000 tonnes (50×10^3 tonnes) of shale per annum.

Until 1998, the Slantsy electric power plant (located close to the Estonian border, 91 miles from St Petersburg) was equipped with oil shale-fired furnaces, but in 1999 its 75 MW plant was converted to use natural gas. The plant continued to process oil shale for oil until June 2003, when the main fuels for the plant changed to coal and natural gas.

In 2002, the Leningradslanets Oil Shale Mining Public Company produced 1.12 million tonnes (1.3×10^6 U.S. tons). In June 2003, all shale mined was delivered to the Estonian Baltic power station with the resultant electricity delivered to UES (Unified Energy System of Russia). However, production ceased at the Leningradslanets Mine on April 1, 2005. Oil shale production restarted on January 15, 2007, with the 50 000 tonnes (55 000 U.S. tons) per month being stored. Leningradslanets exported 40 000 tonnes (44 000 U.S. tons) of oil shale to Estonia between May and August 2009.

2.3.19 Serbia

More than 20 oil shale deposits have been located in Serbia, most in the southern half of the country. The total oil shale resource is estimated to be in the region of 4.8 billion tonnes (5.3×10^9 U.S. tons) with approximately two million barrels of shale oil thought to be recoverable. However, sections of only two of the deposits have received detailed study: (1) Aleksinac in the basin of the same name and (2) Goč-Devotin in the Vlase-Golemo Selo basin.

Viru Keemia Grupp of Estonia has been collaborating with the University of Belgrade to conduct further research and analysis of the oil shale resource. Technical data are not readily available at the time of writing.

2.3.20 Sweden

The oil shale resources underlying mainland Sweden are more correctly referred to as alum shale; black shale is found on two islands lying off the coast of south-eastern Sweden (Andersson et al., 1985). The potential for shale oil production is estimated to be 6.1 billion barrels.

The exploitation of alum shale began as early as 1637 when potassium aluminum sulfate (alum) was extracted for industrial purposes. By the end of the nineteenth century, the alum shale was also being retorted in an effort to produce shale oil. Before and during World War II, Sweden derived oil from its alum shale, but this process had ceased by 1966 when alternative supplies of lower-priced petroleum were available; during the period, 50 million tonnes (55×10^6 U.S. tons) of shale had been mined.

The Swedish alum shale has a high content of various metals including uranium, which was mined between 1950 and 1961. At that time, the available uranium ore was of low grade, but later higher grade ore was found and 50 tonnes of uranium were produced per year between 1965 and 1969. Although the uranium resource is substantial, production ceased in 1989 when world prices decreased and made the exploitation uneconomic.

Sustained commodity price increases in recent years have resulted in a Canadian company, Continental Precious Minerals, conducting a drilling program on the alum shale. The exploration of oil, uranium, and various minerals are all possibilities, and samples are being analyzed by the Estonian Oil Shale Institute.

2.3.21 Syria
Although the existence of oil shale has been known about for the past 60 years, it is only in the recent years of high oil prices that the widely distributed deposits have received more detailed study (Puura et al., 1984).

The most significant and evaluated deposits have been located in the southern Yarmuk Valley, close to the border with Jordan, with the Dar'a deposit having had the most detailed study. Further investigative research and evaluation, particularly in the northern areas of the country, are being undertaken by the General Establishment of Geology and Mineral Resources.

Current unrest at the time of writing makes the future development of Syrian oil shale resources uncertain.

2.3.22 Thailand
Some exploratory drilling by the government was done as early as 1935 near Mae Sot in Tak Province on the Thai-Burmese border. The oil shale beds are relatively thin and the structure of the deposit is complicated by folding and faulting.

Approximately 18.7 billion tonnes (20.6×10^9 U.S. tons) of oil shale have been identified in Tak Province, but to date it has not been economic to exploit these deposits. Proved recoverable reserves of shale oil are put at 810 million tonnes (890×10^6 U.S. tons) (Vanichseni et al., 1988).

Another deposit at Li (Lampoon Province) is small, estimated at 15 million tonnes (16.5×10^6 U.S. tons) of oil shale and yielding 10–45 gallons of shale oil per ton.

The Thai Government has instituted a 4-year project to study the feasibility of developing and utilizing the Mae Sot oil shale deposit. The potential for both direct use (electricity generation) and indirect use (extraction of shale oil) is being evaluated and there is also an investigation of the suitability of using the retort ash in the building industry.

2.3.23 Turkey
Oil shale comprises the second largest potential fossil fuel in Turkey (Altun et al., 2006; Güleç and Önen, 1993; Sener et al., 1995). The main oil shale resources are located in middle and western regions of Anatolia. The amount of proved explored reserves is around 2.22 billion (2.2×10^9) tons, while the total reserves are predicted to be 3–5 billion (3 to 5×10^9) tons. Despite this vast potential, the stated amount cannot be accepted as the amount of commercial reserves. Four major deposits, Himmetoğlu, Seyitömer, Hatildağ, and Beypazari, have been studied in detail and found to vary quite widely in quality.

The deposits vary from 500 to 4500 kcal/kg in calorific value, revealing that each deposit requires a detailed study regarding its possible use (Güleç and Önen, 1993). Numerous studies carried out to recover shale oil have ended with positive but unfeasible results.

However, it is already considered that in general Turkish oil shale would be most profitably used to supplement coal or lignite as a power station fuel, rather than for the recovery of shale oil.

2.3.24 United Kingdom
The oil shale industry was started in Scotland where, in 1694, oil was produced by heating Shropshire oil shale. The direct combustion of oil shale to produce hot water, steam, and, finally, electricity has developed in accordance with the general trends in solid fuel combustion technology. At the beginning of the nineteenth century, industrialized countries

became more interested in obtaining oil and gas from coal pyrolysis (the decomposition or transformation of the kerogen organic matter into hydrocarbons by heat).

Thus, the United Kingdom (specifically Scotland) holds an important place in the historical development of oil shale working. The first oil shale patent—a way to extract and make great quantities of pitch, tarr, and oyle out of a sort of stone (Crown Patent No. 330)—was granted to a group of English entrepreneurs in 1694, and the first substantial shale-oil industry was begun in the Lothians of Scotland (based on Carboniferous oil shales) in 1851 by John Young and others. Production peaked in 1913 when over 3.2 million tons of oil shale was processed. The industry declined slowly but steadily until it closed down in 1962.

No matter what the method of formation, the use of oil shale as a fuel or fuel source is not new and can be traced back to ancient times. But to skip a generation or two, the modern use of oil shale to produce oil dates to Scotland in the mid-nineteenth century (Louw and Addison, 1985). In 1847, Dr. James Young prepared lighting oil, lubricating oil, and wax from coal. Then he moved his operations to Edinburgh where oil shale deposits were found, and in 1850, he patented the process of cracking the oil into its constituent parts. Thus, oil from oil shale was produced in that region from 1857, but the production (albeit small-scale production) was terminated in 1966 because of the availability of cheaper supplies of petroleum crude oil.

Oil shale occurs at a number of other stratigraphical levels in Great Britain, notably in the Devonian Caithness Flags and in the Jurassic Lias, Dun Caan and Brora oils shales, Oxford Clay, and Kimmeridge Clay. Of these, the Kimmeridge Clay has long seemed the most economically interesting prospect, but repeated attempts at commercial exploitation have ended in failure.

The occurrence of oil shale in the cliffs of Kimmeridge Clay at Kimmeridge Bay, Dorset, has been known since the Iron Age. The most famous seam, the Blackstone, has been used locally as a coal substitute and has yielded at various times products ranging from lubricating oil to sanitary deodorizer. During the latter half of the nineteenth century, eight companies were set up to exploit this oil shale, but none was lastingly successful. In each case, the failure was blamed on the unacceptably higher sulfur content (4–8% w/w) of the shale-oil combined with the high cost of working thin seams.

Combustible shales have been recorded in the Kimmeridge Clay throughout its English outcrop. Assessments were made of their potential value as fuel in Dorset and Lincolnshire and Norfolk during the World War I, and although substantial quantities of good quality oil shale were reported to be present, no major industry was developed. They were reassessed by the British Geological Survey in the 1970s as a result of the economic crisis caused by a rapid increase in crude oil prices in 1973.

Oil shale is present throughout the outcrop and subcrop of the Kimmeridge Clay, locally with over 100 seams with potential yields of shale oil ranging from 10 to 90 gallons per ton. However, major economic and environmental problems would need to be solved before they could be worked on a large scale. The seams are thin (most are <6 feet) and separated by barren mudstones that would have to be removed before the oil shale concentrate could be retorted at 500°C (930°F) to yield shale oil. This pyrolysis produces sulfurous gases and large volumes of spent shale, and the shale oils and spent shale can contain low concentrations of carcinogens. In addition, the shale oil would have to be distilled to make them comparable to naturally occurring crude oils that can be used as a refinery feedstock.

The oil shale in the Kimmeridge Clay has the potential to produce millions of tons of shale oil, but only by removing and processing tens of cubic kilometers of material from opencast excavations, an impossible task in a densely populated country such as Britain. Even if this was possible, the energy used in the winning and upgrading processes might be greater than the energy value of the finished product. The oil shale in the Kimmeridge Clay could never, therefore, make a major contribution to the energy supply of the United Kingdom.

2.3.25 United States of America

Deposits of oil shale, ranging from Precambrian to Tertiary age, are present in the United States. The two most important deposits are in the Eocene Green River Formation in Colorado, Wyoming, and Utah (Figure 2.3) and in the Devonian–Mississippian black shales in the eastern United States (Conant and Swanson, 1961; de Witt et al., 1993; Dyni, 2003, 2006; Johnson et al., 2009; Pitman et al., 1989; US DOE, 2007). Oil shale associated with coal deposits of Pennsylvanian age is also

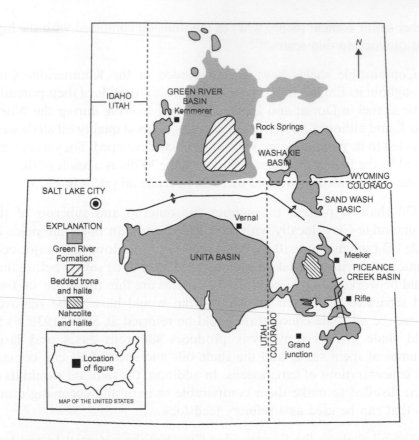

Fig. 2.3. Main Basins of the Green River Formation.

present in the eastern United States. Other deposits are known to be located in Nevada, Montana, Alaska, Kansas, and elsewhere, but these have not been sufficiently explored to be considered as commercial resources (Russell, 1990). Because of their size and grade, most investigations have focused on the Green River and the Devonian–Mississippian deposits.

The largest of the deposits is found in the Eocene Green River Formation in north-western Colorado, north-eastern Utah, and south-western Wyoming. The richest and most easily recoverable deposits are located in the Piceance Creek Basin in western Colorado and the Uinta Basin in eastern Utah. The shale oil can be extracted by surface and in situ methods of retorting: depending upon the methods of mining and processing used, as much as one-third or more of this resource might be recoverable. There is also the Devonian–Mississippian black shale in the eastern United States.

The proven amount of U.S. oil shale resources, the proved amount of shale in place, is put at 3680×10^9 U.S. tons, of which approximately 89% is located in the Green River deposits and 11% in the Devonian black shale.

By assay techniques (Fischer assay being the commonly used method), oil yields vary from approximately 10–50 gallons per ton and, for a few feet in the Mahogany zone, up to about 65 gallons per ton. Oil shale yields more than 25 U.S. gallons per ton and are generally accepted as the most economically attractive and, hence, the most favorable for development.

Recoverable resources of shale oil from the marine black shale in the eastern United States were estimated in 1980 to exceed 400 billion barrels (400×10^9 bbls). These deposits differ significantly in chemical and mineralogical composition from Green River oil shale. Owing to its lower atomic hydrogen or carbon ratio, the organic matter in eastern oil shale yields only about one-third as much oil as Green River oil shale, as determined by conventional Fischer assay analyses. However, when retorted in a hydrogen atmosphere, the oil yield of eastern oil shale increases by as much as two to three times the Fischer assay yield.

Many pilot retorting processes have been tested for short periods. Among the largest was a semi-commercial-size retort operated by Union Oil in the late 1950s, which processed 1100 tons per day of high-grade shale. A pilot plant operated by Tosco (The Oil Shale Corporation) processed 900 tons per day of high-grade shale in the early 1970s. For a shale grade of 37 gallons (150 liters) per ton, these feed rates correspond, respectively, to production of 43 500 gallons (165 000 liters) per day and 357 000 gallons (135 000 liters) per day of crude shale oil.

Because of the abundance and geographic concentration of the known resource, oil shale has been recognized in the United States as a potentially valuable energy resource, since as early as 1859, the same year Colonel Drake completed his first oil well in Titusville, Pennsylvania. Oil distilled from oil shale was first burnt for horticultural purposes in the nineteenth century, and early products derived from shale oil included kerosene and lamp oil, paraffin, fuel oil, lubricating oil and grease, naphtha, illuminating gas, and ammonium sulfate fertilizer. However, although the U.S. shale oil industry was a viable part of the economy before Drake's strike at Titusville, Pennsylvania, in 1859, the industry struggled after the Titusville discovery and practically disappeared within a few years.

However, in the early 1900s, more detailed investigations were made, and in 1912, the U.S. government established the Office of Naval Petroleum and Oil Shale Reserves. The oil shale reserves were seen as a possible emergency source of fuel for the military, particularly the United States Navy, which had, at the beginning of the twentieth century, converted its ships from coal to fuel oil. The nation's economy was transformed by gasoline-fueled automobiles and diesel-fueled trucks and trains, and concerns have been raised about assuring adequate supplies of liquid fuels at affordable prices to meet the growing needs of the nation and its consumers.

The abundance of oil shale resources in the United States were initially eyed as a major source for these fuels. Numerous commercial entities sought to develop oil shale resources. The Mineral Leasing Act of 1920 made petroleum and oil shale resources on Federal lands available for development under the terms of federal mineral leases. Soon, however, discoveries of more economically producible and refinable liquid crude oil in commercial quantities caused interest in oil shale to decline.

Interest resumed after World War II, when military fuel demand, domestic fuel rationing and rising fuel prices made the economic and strategic importance of the oil shale resource more apparent (US DOE, 2004a, b, c). After the war, the booming post-war economy drove the demand for fuels even higher. Public and private research and development efforts commenced, including the 1946 United States Bureau of Mines Anvil Point, Colorado oil shale demonstration project. Significant investments were made to define and develop the resource and to develop commercially viable technologies and processes to mine, produce, retort, and upgrade oil shale into viable refinery feedstocks and by-products. Once again, however, major crude oil discoveries in the continental 48 States, off-shore, and in Alaska, as well as other parts of the world reduced the foreseeable need for shale oil, and interest and associated activities again diminished. Lower-48 United States crude oil reserves peaked in 1959 and lower-48 production peaked in 1970.

By 1970, oil discoveries were slowing, demand was rising, and crude oil imports, largely from Middle Eastern states, were rising to meet demand. Global oil prices, while still relatively low, were also rising, reflecting the changing market conditions. On-going oil shale research and testing projects were re-energized and new projects were envisioned by numerous energy companies seeking alternative fuel feedstocks. These efforts were significantly amplified by the impacts of the 1973 Arab Oil Embargo, which demonstrated the nation's vulnerability to oil import supply disruptions, and were underscored by a new supply disruption associated with the 1979 Iranian Revolution.

By 1982, however, technology advances and new discoveries of off-shore oil resources in the North Sea and elsewhere provided new and diverse sources for United States oil imports and dampened global energy prices. Global political shifts promised to open previously restricted provinces to oil and gas exploration and led economists and other experts to predict a long future of relatively low and stable oil prices. Despite significant investments by United States energy companies, numerous variations and advances in mining, restoration, retorting, and in situ processes, the costs of oil shale production relative to foreseeable oil prices, made continuation of most commercial efforts impractical. In addition, the collapse of world oil prices in 1984 did not help the oil shale development cause.

Thus, the prospects for oil shale development in the United States remain uncertain (Bartis et al., 2005). The estimated cost of surface retorting remains high—for surface retorting, it may be inappropriate to contemplate near-term commercial efforts. Meanwhile, the technical groundwork may be in place for a fundamental shift in oil shale economics. Advances in thermally conductive in situ conversion may cause shale-derived oil to be competitive with crude oil. If this becomes the case, oil shale development could soon occupy a very prominent position in the national energy agenda (Bartis et al., 2005).

REFERENCES

Alali, J., 2006. Jordan oil shale, availability, distribution, and investment opportunity. Paper No. RTOS-A117. In: Proceedings of International Conference on Oils Shale: Recent Trends in Oil Shale, Amman, Jordan, November 7–9, 2006.

Altun, N.E., Hiçyilmaz, C., Hwang, J.-Y., Suat Bağci, A.S., Kök, M.V., 2006. Oil shales in the World and Turkey—reserves, current situation and future prospects: a review. Oil Shale 23 (3), 211–227.

Andersson, A., Dahlman, B., Gee, D.G., Snäll, S., 1985. The Scandinavian Alum Shales: Overages Geologiska Undersökning, Avhandlingar Och Uppsatser I A4. Ser. Ca. 56, 50.

Australian Government, 2010. Australina Energy Resource Assessment. Geoscience Australia, Department of Resources, Energy and Tourism, Government of Australia, Canberra, Australian Capital Territory, Australia (Chapter 3).

Baker, J.D., Hook, C.O., 1979. Chinese and Estonian oil shale. In: Proceeding of 12th Oil Shale Symposium. Colorado School of Mines, Golden, CO, pp. 26–31.

Ball, F.D., Macauley, G., 1988. The Geology of New Brunswick Oil Shales, Eastern Canada. In: Proceedings of International Conference on Oil Shale and Shale Oil, Beijing, China, pp. 34–41.

Bartis, J.T., LaTourette, T., Dixon, L., Peterson, D.J., Cecchine, G., 2005. Oil Shale Development in the United States. Report MG-414-NETL. RAND Co., Santa Monica, CA.

Bauert, H., 1994. The baltic oil shale basin—an overview. In: Proceedings of 1993 Eastern Oil Shale Symposium. Institute for Mining and Minerals Research, University of Kentucky, Lexington, KY, pp. 411–421.

Bekri, O., 1992. Possibilities for oil shale development in Morocco. Energeia 3 (5), 1–2.

Bouchta, R., 1984. Valorization Studies of the Moroccan Oil Shales. In: Office Nationale de Researches et Exploitations Petrolieres Agdal, Rabat, Morocco.

Brendow, K., 2003. Global oil shale issues and perspectives. Oil Shale 20 (1), 81–92.

Brendow, K., 2009. Oil shale—a local asset under global constraint. Oil Shale 26 (3), 357–372.

Bsieso, M.S., 2003. Jordan's experience in oil shale studies employing different technologies. Oil Shale 20 (3), 360–370.

Conant, L.C., Swanson, V.E., 1961. Chattanooga Shale and Related Rocks of Central Tennessee and Nearby Areas. US Professional Paper No. 357. US Geological Survey, US Department of the Interior, Washington, DC.

Cook, A.C., Sherwood, N.R., 1989. The oil shales of Eastern Australia. In: Proceedings of 1988 Eastern Oil Shale Symposium. Institute for Mining and Minerals Research, University of Kentucky, Lexington, KY, pp. 185–196.

Crisp, P.T., Ellis, J., Hutton, A.C., Korth, J., Martin, F.A., Saxby, J.D., 1987. Australian Oil Shales—A Compendium of Geological and Chemical Data: North Ryde, New South Wales, Australia. Division of Fossil Fuels, CSIRO Institue of Energy and Earth Sciences, Clayton, South Victoria, Australia.

Culbertson, W.C., Pitman, J.K., 1973. Oil Shale in United States Mineral Resources. Paper No. 820. United States Geological Survey, Washington, DC.

Culbertson, W.C., Smith, J.W., Trudell, L.G., 1980. Oil Shale Resources and Geology of the Green River Formation in the Green River Basin, Wyoming. Report No. LETC/RI-80/6. US Department of Energy, Washington, DC.

Davies, G.R., Nassichuk, W.W., 1988. An early Carboniferous (Viséan) lacustrine oil shale in the Canadian Arctic Archipelago. Bull. Am. Assoc. Petrol. Geol. 72, 8–20.

de Witt, Wallace Jr., Roen, J.B., Wallace, L.G., 1993. Stratigraphy of Devonian Black Shales and Associated Rocks in the Appalachian Basin. Petroleum Geology of the Devonian and Mississippian Black Shale of Eastern North America. Bulletin No. 1909. United States Geological Survey, Washington, DC (Chapter B), pp. B1–B57.

Duncan, D.C., Swanson, V.E., 1965. Organic-Rich Shale of the United States and World Land Areas. Circular No. 523. United States Geological Survey, Washington, DC.

Dyni, J.R., 2003. Geology and resources of some world oil-shale deposits. Oil Shale 20 (3), 193–252.

Dyni, J.R., 2006. Geology and Resources of Some World Oil Shale Deposits. Report of Investigations 2005-5295. United States Geological Survey, Reston, VA.

Ekweozor, C.M., Unomah, G.I., 1990. First discovery of oil shale in the Benue Trough, Nigeria. Fuel 69, 503–508.

Fainberg, V., Hetsroni, G., 1996. Research and development in oil shale combustion and processing in Israel. Oil Shale 13, 87–99.

Fang, C., Zheng, D., Liu, D., 2008. Main problems in development and utilization of oil shale and status of the in situ conversion process technology in China. In: Proceedings of 28th Oil Shale Symposium, October 13–15, 2008. Colorado School of Mines, Golden, CO.

Güleç, K., Önen, A., 1993. Turkish oil shales: reserves, characterization and utilization. In: Proceedings of 1992 Eastern Oil Shale Symposium. Institute for Mining and Minerals Research, University of Kentucky, Lexington, KY, pp. 12–24.

Hamarneh, Y., 1998. Oil Shale Resources Development in Jordan. Natural Resources Authority, Hashemite Kingdom of Jordan, Amman, Jordan.

Hyde, R.S., 1984. Oil Shales near Deer Lake, Newfoundland. Open-File Report No. OF 1114. Geological Survey of Canada, Ottawa, ON, Canada.

Jaber, J.O., Probert, S.D., Badr, O., 1997. Prospects for the exploitation of Jordanian oil shale. Oil Shale 14, 565–578.

Johnson, H.R., Crawford, P.M., Bunger, J.W., 2004. Strategic Significance of America's Oil Shale Resource. Volume II Oil Shale Resources. Technology and Economics. Office of Deputy Assistant Secretary for Petroleum Reserves. Office of Naval Petroleum and Oil Shale Reserves United States Department of Energy, Washington, DC.

Johnson, R.C., Mercier, T.J., Brownfield, M.E., Pantea, M.P., Self, J.G., 2009. Assessment of in-place oil shale resources of the Green River Formation, Piceance Basin, western Colorado. Fact Sheet 2009–3012. U.S. Geological Survey, Washington, DC.

Kalkreuth, W.D., Macauley, G., 1987. Organic petrology and geochemical (Rock-Eval) studies on oil shales and coals from the Pictou and Antigonish Areas, Nova Scotia, Canada. Bull. Can. Petrol. Geol. 35, 263–295.

Kashirskii, V., 1996. Problems of the development of the Russian oil shale industry. Oil Shale 13, 3–5.

Kattai, V., Lokk, U., 1998. Historical review of the kukersite oil shale exploration in Estonia. Oil Shale 15 (2), 102–110.

Liang, Y., 2006. Current Status of the Oil Shale Industry in Fushun, China. Paper No. RTOS-A106. In: Proceedings of International Oil Shale Conference: Recent Trends in Oil Shale, Amman, Jordan, November 7–9, 2006.

Lippmaa, E., Maramäe, E., 1999. Dictyonema shale and uranium processing at Sillamäe. Oil Shale 16, 291–301.

Lippmaa, E., Maramäe, E., 2000. Uranium production from the local dictyonema shale in Northeast Estonia. Oil Shale 17, 387–394.

Lippmaa, E., Maramäe, E., 2001. Extraction of uranium from local Dictyonema shale at Sillamäe in 1948–1952. Oil Shale 18, 259–271.

Loog, A., Aruväli, J., Petersell, V., 1996. The nature of potassium in Tremadocian Dictyonema shale (Estonia). Oil Shale 13, 341–350.

Louw, S.J., Addison, J., 1985. Studies of the Scottish oil shale industry. Research Report TM/85/02, DOF/ER/60036IOM/TM/85/2. US Department of Energy Project DE-ACO2 – 82ER60036. US Department of Energy, Washington, DC.

Macauley, G., 1981. Geology of the Oil Shale Deposits of Canada. Open-File Report OF-754, Geological Survey of Canada, Ottawa, ON, Canada.

Macauley, G., 1984a. Cretaceous Oil Shale Potential of the Prairie Provinces of Canada Open-File Report OF-977. Geological Survey of Canada, Ottawa, ON, Canada.

Macauley, G., 1984b. Cretaceous Oil Shale Potential in Saskatchewan. Special Publication 7, Saskatchewan Geological Society, Saskatoon, Saskatchewan. pp. 255–269.

Macauley, G., Snowdon, L.R., Ball, F.D., 1985. Geochemistry and geological factors governing exploitation of selected Canadian oil shale deposits. Paper 85-13. Geological Survey of Canada, Ottawa, ON, Canada.

Matthews, R.D., Janka, J.C., Dennison, J.M., 1980. Devonian Oil Shale of the Eastern United States, a Major American Energy Resource. In: American Association of Petroleum Geologists Meeting, Evansville, IN.

Matthews, R.D., 1983. The Devonian-Mississippian oil shale resource of the United States. In: Proceedings of 16th Oil Shale Symposium, Colorado School of Mines, Golden, CO, pp. 14–25.

Minster, T., 1994. The role of oil shale in the Israeli energy balance. Energia 5 (5), 4–6.

Noon, T.A., 1984. Oil shale resources in Queensland. In: Proceedings of Second Australian Workshop on Oil Shale: Sutherland, NSW, Australia, CSIRO Division of Energy Chemistry, Sutherland New South Wales, Australia.

Ots, A., 2007. Estonian oil shale properties and utilization in power plants. Energetika 53 (2), 8–18.

Ozimic, S., Saxby, J.D., 1983. Oil Shale Methodology—An Examination of the Toolebuc Formation and the Laterally Contiguous Time Equivalent Units, Eromanga and Carpenteria Basins. NERDDC Project 78/2616: Australian Bureau of Mineral Resources and CSIRO, North Ryde, New South Wales, Australia.

Padula, V.T., 1969. Oil shale of Permian Iratí Formation, Brazil. Bull. Am. Assoc. Petrol. Geol. 53, 591–602.

Pitman, J.K., Wahl Pierce, F., Grundy, W.D., 1989. Thickness, Oil-Yield, and Kriged Resource Estimates for the Eocene Green River Formation, Piceance Creek Basin, Colorado. Chart No. OC-132. US Geological Survey Oil, Washington, DC.

Puura, V., Martins, A., Baalbaki, K., Al-Khatib, K., 1984. Occurrence of oil shales in the South of the Syrian Arab Republic (SAR). Oil Shale 1, 333–340.

Qian, J., Wang, J., Li, S., 2003. Oil shale development in China. Oil Shale 20 (3), 356–359.

Reinsalu, E., 1998. Criteria and size of Estonian oil shale reserves. Oil Shale 15 (2), 111–133.

Russell, P.L., 1990. Oil Shales of the World, Their Origin, Occurrence and Exploitation. Pergamon Press, New York.

Schmidt, S.J., 2003. New directions for shale oil: path to a secure new oil supply well into this century (on the example of Australia). Oil Shale 20 (3), 333–346.

Scouten, C., 1990. Fuel Science and Technology Handbook. In: Speight, J.G. (Ed.), Part IV: Oil Shale. Chapters 25–31. Marcel Dekker Inc., New York.

Sener, M., Senguler, I., Kok, M.V., 1995. Geological considerations for the economic evaluation of oil shale deposits in Turkey. Fuel 74, 999–1003.

Sherwood, N.R., Cook, A.C., 1983. Petrology of organic matter in the Toolebuc formation oil shales. In: Proceedings of First Australian Workshop on Oil Shale, May 18–19, 1983, CSIRO Division of Energy Chemistry, Sutherland, New South Wales, Australia, pp. 35–38.

Shi, G-Q., 1988. Shale oil industry in Maoming. In: Proceedings of International Conference on Oil Shale and Shale Oil, Beijing, pp. 670–678.

Smith, W.D., Naylor, R.D., 1990. Oil Shale Resources of Nova Scotia. Nova Scotia Economic Geology Series Report 90-3. Department of Mines and Energy, Halifax, Nova Scotia.

Speight, J.G., 2007. The Chemistry and Technology of Petroleum, fourth ed. CRC-Taylor and Francis Group, Boca Raton, FL.

Speight, J.G., 2008. Synthetic Fuels Handbook: Properties, Processes, and Performance. McGraw-Hill, New York.

Speight, J.G., 2009. Enhanced Recovery Methods for Heavy Oil and Tar Sands. Gulf Publishing Company, Houston, TX.

Speight, J.G., 2011. An Introduction to Petroleum Technology, Economics, and Politics. Scrivener Publishing, Salem, MA.

Speight, J.G., 2013. The Chemistry and Technology of Petroleum, third ed. CRC-Taylor and Francis Group, Boca Raton, FL.

US DOE, 2004a. Strategic Significance of America's Oil Shale Reserves, I. Assessment of Strategic Issues. United States Department of Energy, Washington, DC. http://www.fe.doe.gov/programs/reserves/publications

US DOE, 2004b. Strategic Significance of America's Oil Shale Reserves, II. Oil Shale Resources, Technology, and Economics. United States Department of Energy, Washington, DC. http://www.fe.doe.gov/programs/reserves/publications

US DOE, 2004c. America's Oil Shale: A Roadmap for Federal Decision Making; USDOE Office of US Naval Petroleum and Oil Shale Reserves. United States Department of Energy, Washington, DC. http://www.fe.doe.gov/programs/reserves/publications

US DOE, 2007. Secure Fuels from Domestic Resources, The Continuing Evolution of America's Oil Shale and Tar Sands Industries: Profiles of Companies Engaged in Domestic Oil Shale and Tar Sands Resource and Technology Development. Office of Naval Petroleum and Oil Shale Reserves, Office of Petroleum Reserves, US Department of Energy, Washington, DC.

Vanichseni, S., Silapabunleng, K., Chongvisal, V., Prasertdham, P., 1988. Fluidized bed combustion of Thai oil shale. In: Proceedings of International Conference on Oil Shale and Shale Oil, Chemical Industry Press, Beijing, China, pp. 514–526.

Zhou, C., 1995. General description of Fushun oil shale retorting factory in China. Oil Shale 13, 7–11.

CHAPTER 3

Kerogen

3.1 INTRODUCTION

The name *kerogen* is also generally used for organic matter in sedimentary rocks that is insoluble in common organic and inorganic solvents. Thus, the term *kerogen* is used throughout this chapter to mean the carbonaceous material that occurs in sedimentary rocks, carbonaceous shale, and oil shale. This carbonaceous material is, for the most part, insoluble in common organic solvents. A soluble fraction, *bitumen*, coexists with the kerogen. This bitumen is not to be confused with the material found in tar sand deposits (Speight, 2007, 2008, 2009). However, like many naturally occurring organic materials, kerogen does yield a hydrocarbonaceous oil when heated to temperatures sufficiently high (typically >300°C, 570°F) to cause thermal decomposition with simultaneous removal of distillate.

The precise details regarding the perennial issues related to the origin and accumulation of kerogen have yet to be fully answered. Therefore, in any text dealing with the science and technology of oil shale, there must, of necessity, be a section dealing with kerogen, but it is not the goal of this chapter to deal with the various intimacies of kerogen structure. The chapter is to assist the reader to understand kerogen and its place as a naturally occurring organic material (Durand, 1980; Scouten, 1990; Tissot and Welte, 1978; Vandenbroucke, 2003).

Kerogen is the naturally occurring, solid, insoluble organic matter that occurs in source rocks and can yield oil upon heating. Typical organic constituents of kerogen are algae and woody plant material. Kerogen has a high molecular weight relative to tar sand bitumen and is generally insoluble in typical organic solvents (Speight, 2007, 2009). It has been conveniently divided into four types: (1) Type I, which consists mainly of algal and amorphous constituents; (2) Type II, which consists of mixed terrestrial and marine source material; (3) Type III, which consists of woody terrestrial source material; and (4) Type IV, which consists mostly of decomposed organic matter in the form of polycyclic aromatic hydrocarbons and has a low (<0.5) hydrogen-to-carbon (H/C) atomic ratio.

3.2 COMPOSITION AND PROPERTIES

Kerogen is a solid, waxy, organic substance that is formed when pressure and heat from the Earth act on the remains of plants and animals.

In very general terms, the hydrogen content of kerogen falls between that of petroleum and that of coal (Speight, 2007, 2013), but this varies considerably with the source so that a range of values is found. This has been suggested as reflecting an overlap between terrestrial and aquatic origin. In fact, high lipid content, consistent with the occurrence of aquatic plants in the source material, appears to be diminished in kerogen by lignin of terrestrial origin (Scouten, 1990, and references cited therein). In fact, kerogen is best represented as a macromolecule that contains considerable amounts of carbon and hydrogen (Scouten, 1990). Furthermore, it is the macromolecular and heteroatomic nature of kerogen, with up to 400 heteroatoms (nitrogen plus oxygen plus sulfur) for every 1000 carbon atoms occurring as an integral part of the macromolecule, that classifies kerogen as a naturally occurring heteroatomic material.

Given geological time, it is believed that kerogen converts to various liquid and gaseous *hydrocarbons* at a depth of approximately 4.5 miles or more (approximately 7 km) and a temperature between 50°C and 100°C (122°F and 212°F) (USGS, 1995), which has been assigned to the presence of the *thermal gradient*.

Briefly, the *geothermal gradient* is the variation of temperature with depth in subterranean formations of the Earth (Chapter 1). Although this varies from place to place, it is generally in the order of 22°F per 1000 feet of depth or 12°C per 1000 feet of depth, that is, 0.022°F per foot of depth or 0.012°C per foot of depth. This would require a depth in the order of 25 000 to attain temperatures of 300°C (570°F).

Since geological time cannot be considered by the laboratory scientists, much laboratory work has focused on increasing the temperature to increase the reaction rate to study the thermal evolution of kerogen. However, the application of high temperatures (>250°C, <480°F) not only increases the rate of reaction (thereby making up for the lack of geological time) but can also change the *chemistry* of the reaction. Furthermore, introduction of a pseudo-activation energy in which the activation energy of the kerogen conversion reactions is reduced leaves much to be desired because of the assumption required to develop this pseudo-activation energy equation(s). Nor is it valid to use a fixed set of kinetic parameters within each of these groups (Whelan and Farrington, 1992). Suffice it to state that the thermal evolution of kerogen is unknown and our knowledge of the role of kerogen in petroleum formation is at best, highly speculative.

In terms of the formation of kerogen, it can only be assumed that the similar types of plant debris that went into the formation of petroleum and coal may have played a role in the formation of kerogen (Erdman, 1981; Scouten, 1990; Speight, 2007, 2013; Tissot et al., 1978).

Once incorporated into sediments, the organic matter is buried under increasing depths as deposition of the mineral matter continues (sedimentation). Within the sediment, the physicochemical and biological environment is then gradually modified by the following events: (1) compaction; (2) decrease in water content; (3) cessation of bacterial activity; (4) transformation of the mineral phase; and (5) to some extent, but largely unknown, an increase in temperature. Under these conditions, the skeletal structures of the plant materials could be preserved to a significant degree. Should this be the case, there is the distinct possibility that oil and kerogen are produced from the organic material by simultaneous or closely consecutive processes (Figure 3.1). There is also the theory that lignin derivatives do not usually form oil but are more likely to produce coal (Speight, 2007, 2013). However, it must be remembered that these statements are theories, and proof, other than the questionable proof obtained by laboratory experiments, is difficult (perhaps even impossible) to obtain.

The Van Krevelen diagram (a plot of the atomic hydrogen-carbon ratio versus the atomic oxygen-carbon ratio) (Figure 3.2), derived from the elemental analysis of kerogen and coal, is a practical means of studying kerogen composition and properties (Speight, 2007, 2013).

The position of kerogen in the H/C–O/C diagram is related to the total quantity of hydrocarbons, which in turn is a function of the relative amounts of aromatic hydrocarbon structures. The data for kerogen analysis in the H/C–O/C diagram can be considered to describe the evolutionary path for kerogen from different precursors. Analysis of the

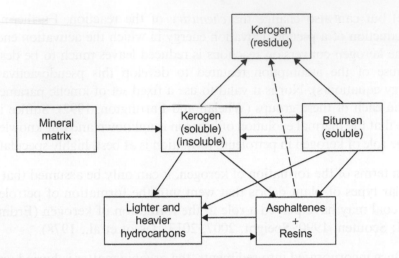

Fig. 3.1. Hypothetical representation of petroleum formation from kerogen.

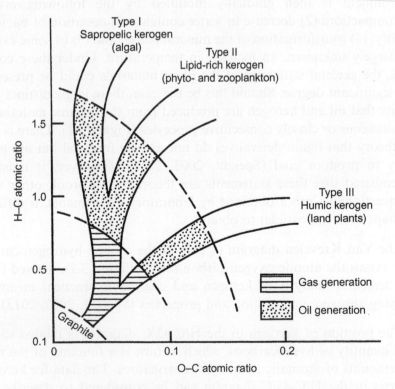

Fig. 3.2. Hypothetical evolutionary pathways (atomic H/C ratio vs. atomic O/C ratio) of kerogen.

minor elements, sulfur and nitrogen, is much more difficult to simulate and may require a more detailed framework. The general characteristics of the four types of kerogen are as follows:

Type I kerogen

- Alginite
- Hydrogen/carbon atomic ratio >1.25.
- Oxygen/carbon atomic ratio <0.15.
- Tendency to readily produce liquid hydrocarbons.
- Derived principally from lacustrine algae.
- Has few cyclic or aromatic structures.
- Formed mainly from proteins and lipids.

Type II kerogen

- Hydrogen/carbon atomic ratio <1.25.
- Oxygen/carbon atomic ratio 0.03 to 0.18.
- Tends to produce a mix of gas and oil.
- Several types: exinite, cutinite, resinite, and liptinite.
- Exinite is formed from pollen and spores.
- Cutinite is formed from terrestrial plant cuticle.
- Resinite is terrestrial plant resins, animal decomposition resins.
- Liptinite is formed from terrestrial plant lipids and marine algae.

Type III kerogen

- Hydrogen/carbon atomic ratio <1.0.
- Oxygen/carbon atomic ratio 0.03 to 0.3.
- Material is thick, resembling wood or coal.
- Tends to produce coal and gas.
- Has very low hydrogen content because of the extensive ring and aromatic systems.
- Formed from terrestrial plant matter that is lacking in lipids or waxy matter; forms from cellulose, the carbohydrate polymer that forms the rigid structure of terrestrial plants, lignin, another carbohydrate polymer (polysaccharide) that binds the strings of cellulose together, and terpenes and phenolic compounds in the plant.

Type IV kerogen (residue)

- Hydrogen/carbon atomic ratio <0.5.
- Contains mostly decomposed organic matter in the form of polycyclic aromatic hydrocarbons.
- Little or no potential to produce hydrocarbons.

Type I kerogen is rich in lipid-derived aliphatic chains and has a relatively low content of polynuclear aromatic systems and heteroatomic systems. The initial atomic H/C ratio is high (1.5 or more), and the atomic O/C ratio is generally low (0.1 or less). This type of kerogen is generally of lacustrine origin. Organic sources for the type I kerogen include the lipid-rich products of *algal blooms* and the finely divided and extensively reworked lipid-rich biomass deposited in stable stratified lakes.

Type II kerogen is characteristic of the marine oil shales. The organic matter in this type of kerogen is usually derived from a mixture of zooplankton, phytoplankton, and bacterial remains that were deposited in a reducing environment. Atomic H/C ratios are generally lower than for type I kerogen, but the O/C atomic ratios are generally higher for type II kerogen than for type I kerogen. Organic sulfur levels are also generally higher in the type II kerogen. The oil-generating potential of type II kerogen is generally lower than that of type I kerogen (i.e., less of the organic material is liberated as oil upon heating a type II kerogen at the same level of maturation).

Type III kerogen is characteristic of coals and coaly shales. Easily identified fossilized plants and plant fragments are common, indicating that this type of kerogen is derived from woody terrestrial material. These materials have relatively low atomic H/C ratios (usually <1.0) and relatively high atomic O/C ratios (>0.2). Aromatic and heteroaromatic contents are high, and ether units (especially of the diaryl ethers) are important, as might be anticipated for a lignin-derived material. Oil-generating potentials are low, but gas-generating potentials are high.

At the beginning of the type I path, the kerogen types have a strongly aliphatic nature; at the beginning of the type III path, the kerogen consists largely of aromatic structures that carry oxygen functions. At the beginning of the type II path, and in general for intermediate paths between type I and type III, elemental analysis supplies little information about the chemical structure.

3.3 ISOLATION

The first step in any study of the behavior and structure of kerogen has generally been the isolation of a kerogen concentrate (Durand, 1984; Forsman, 1963; Forsman and Hunt, 1958; Robinson, 1969; Saxby, 1976). A variety of methods can be employed to isolate fractions of organic material without altering the structure of the native kerogen. There are also those methods intended for degradation of the organic material in a controlled manner. The terminology of the material isolated by such methods is often based on the method employed. Therefore, an understanding of these methods assists in understanding the terminology. For example, a particular method may result in the generation of hydrocarbon products as well as more complex products that are heteroatomic or high in molecular weight.

Physical methods to produce an organic-rich kerogen concentrate are of interest because exposure of the kerogen to strong acid or base is avoided, thereby lessening the chance of chemical alteration. Such methods generally involve the potential for contamination of the kerogen with materials used to effect the separation. However, in many cases, the potential impact of such contaminants can be limited by using only one or a small number of known and easily identified chemical species. Among the more important physical methods for kerogen concentration are sink-float, oil agglomeration, and froth flotation methods (Hubbard et al., 1952; Vadovic, 1983).

By far the most common technique for kerogen isolation involves acid demineralization of the shale to produce the kerogen concentrate. To dissolve the mineral matrix, a series of successive treatments with a hydrochloric acid-hydrofluoric acid mixture (at approximately 65°C, 150°F) is employed (Durand, 1984). On the other hand, demineralization with a reduced chance of organic alteration has been achieved by carrying out the treatment at a lower temperature (20°C, 70°F) and for shorter times (Scouten et al., 1987). The use of base to dissolve silicates has also been investigated followed by an acid treatment to dissolve carbonates (McCollum and Wolff, 1990).

The isolation of kerogen from mineral matrices also depends on the extent of the interactions between the kerogen and the various minerals. From the results of model compound-model mineral tests, interactions between acid clay minerals and nitrogen-containing organic compounds have been identified as being much stronger than other likely candidates for kerogen-mineral interactions (Siskin et al., 1987a, 1987b). The importance of this finding led to the use of differential wetting, a phenomenon typically associated with physical separation methods, as critical to the success of kerogen separation. Thus, efficient kerogen recovery can be achieved by adding an organic solvent that wets and swells the kerogen, thereby diminishing some of the nitrogen-mineral interactions and aiding the physical sink-float separation. Thus, both chemical and physical aspects are important for the production under mild conditions of a kerogen concentrate with, presumably, minimal structural alterations.

The laboratory sink-float methods offer mild conditions to minimize chemical alteration and can produce a kerogen concentrate with low ash content. There are disadvantages to this technique, however, which include (1) rejection of organic compounds (leading to low recovery of the kerogen) and (2) the possibility of kerogen fractionation along with mineral rejection. The oil agglomeration method relies on selective wetting of kerogen particles by an oily paste material such as hexadecane (see also Himus and Basak, 1949; Robinson, 1969; Smith and Higby, 1960).

3.4 METHODS FOR PROBING KEROGEN STRUCTURE

3.4.1 Ultimate (Elemental) Analysis

Although not strictly a method for probing the structure of kerogen, elemental analysis offers valuable information about its atomic constituents. The elemental analysis of kerogen is a method for characterizing the origin and evolution of sedimentary organic matter. Elemental analysis also establishes a framework within which other physicochemical methods can be used more effectively.

The Van Krevelen diagram (a plot of the atomic hydrogen–carbon ratio versus the atomic oxygen–carbon ratio) (Figure 3.2), derived from the elemental analysis of kerogen and coal, is a very practical means of studying kerogen composition. Oil and gas are believed to be formed during the evolutionary path for kerogen from different precursors.

Inferences regarding kerogen structure are drawn from the results of bitumen (kerogen extract) analyses. The conclusions are generally based on the premise that the bitumen is analogous to the original organic matter, that is, the bitumen represents units of the precursor that did not become bound into the insoluble three-dimensional macromolecular network of the kerogen. It is also assumed that the bitumen is representative of units of the kerogen structure that have been cleaved more or less intact, with little or no structural alteration, from the kerogen by thermal treatment.

Many different compound types have been identified (by extraction procedures) as part of the kerogen matrix but their mode of inclusion in kerogen remains open to speculation. For example, these compounds include paraffins (Anderson et al., 1969; Cummins and Robinson, 1964), steranes (Anderson et al, 1969), cycloalkanes (Anders and Robinson, 1971), aromatics, and polar compounds (Anders et al., 1975). Kerogen generally contains only a small amount of bitumen (<15% w/w of the total organic matter), but this usually has a higher hydrogen content than the corresponding kerogen. This corresponds to a lower proportion of aromatics as well as nitrogen-, oxygen-, and sulfur-containing compounds. This is an obvious limitation to the usefulness of structural inferences drawn from the bitumen composition.

The volatile oil generally represents a much larger fraction of the original organic material (usually 50% or more of the available organic carbon). The methods employed for analysis of the product oils are similar to those used for petroleum (Fenton et al., 1981; Holmes and Thompson, 1981; Regtop et al., 1982; Uden et al., 1978; Williams and Douglas, 1981). As a consequence of their thermal treatment, the volatile oils are usually richer in aromatics and olefins than the starting kerogen but are relatively deficient in nitrogen-containing and sulfur-containing compounds. As for petroleum coking (Speight, 2007), the nitrogen- and sulfur-containing species are concentrated in the non-volatile char. The volatile oils are a greater reflection of the thermal treatment used to produce the oil and, consequently, and are not too reliable in terms of an accurate picture of kerogen structure. Thus, although many inferences about kerogen structure have been drawn from bitumen and oil analyses, their limitations must be recognized.

3.4.2 Functional Group Analysis

Attempts to characterize the oxygen functional groups in kerogen have focused on acid demineralization (successive treatments with hydrochloric acid and hydrofluoric acid) to prepare a kerogen concentrate. The concentrate has then been treated by wet chemical methods to determine the distribution of oxygen functional groups (Fester and Robinson, 1966; Robinson and Dineen, 1967).

3.4.3 Oxidation

Oxidative degradation, one of the primary methods of structural determination used in natural product chemistry, has also been employed to examine kerogen structure (Vitorovic, 1980). Alkaline permanganate and chromic acid have been the two most widely used oxidants, although ozone, periodate, nitric acid, perchloric acid, air or oxygen, hydrogen peroxide, and electrochemical oxidation (among many other reagents) have also been used.

Alkaline permanganate oxidation of kerogen has been carried out in two very different ways. Older work generally involved use of the *carbon balance* method developed for studies of coal. The products of this exhaustive oxidation are carbon dioxide, oxalic acid ($HO_2C\text{-}CO_2H$, from aromatic rings), nonvolatile benzene polycarboxylic acids, and unoxidized organic carbon. However, because aliphatic material is oxidized mainly to carbon dioxide, this method is not well suited to probe the structure of kerogen, as it is highly aliphatic. This led to the development of stepwise procedures to give products that retain more structural information about the starting kerogen.

The careful development of the stepwise alkaline permanganate method represented attempts to minimize unwanted secondary oxidation of the first-formed product by adding the oxidant—potassium permanganate in aqueous potassium hydroxide ($KMnO_4$ in 1% aqueous KOH) in small portions. In some cases, the acids obtained from stepwise oxidation proved to be of such high molecular weight that they precipitated upon acidification, were insoluble in ether, and were difficult to characterize. In these cases, the precipitated acids were subjected to further stepwise oxidation to produce the desired ether-soluble acids of lower molecular weight.

In general terms, alkaline permanganate oxidizes alkylbenzenes, alkylthiophenes, and alkylpyridines (but not alkylfurans) to the corresponding carboxylic, acids. This is not true when the aromatic ring bears an electron-donating group (e.g., –OH, –OR, or –NH$_2$). In such cases, degradation of the aromatic portion is usually rapid. Condensed aromatics are also attacked and benzene polycarboxylic acids are produced. In addition, caution is advised since benzene itself also reacts (slowly) in hot alkaline permanganate solutions. Olefins are rapidly converted into the corresponding glycol, which are then cleared to carboxylic acids.

$$RCH=CHR^1 \rightarrow RCH(OH)CH(OH)R^1 \rightarrow RCO_2H + RCO_2H$$

Cyclic olefins yield dicarboxylic acids. Enolizable ketones are also cleaved, presumably via the enol.

Tertiary and benzylic carbon-hydrogen groups are reacted to afford tertiary alcohols. In simple alkyl systems, the presence of an alcohol group markedly accelerates the rate of this reaction. Primary and secondary alcohols are oxidized to the corresponding acids and ketones, and alkaline permanganate oxidation degrades the porphyrin nucleus, giving pyrrole-2, 4-dicarboxylic acid derivatives under mild conditions. The porphyrin side chains $-CH_3$, $-CH_2CH_3$, $-CH_2CH_2CO_2H$, $-COCH_3$, and $-CH(OH)CH_3$ persist in the degradation products, but the $-CH=CH_2$ and $-CHO$ side chains are both oxidized to $-CO_2H$.

The structural information obtained from the oxidation of kerogen by chromic acid (and other chromium-containing oxidants) is usually similar to that obtained with alkaline permanganate (Lee, 1980; Vitorovic, 1980). However, for any particular technique, the recovery of organic carbon in the oxidation products may be as low as 10% of the original carbon. Consequently, the alkaline permanganate procedure is often considered superior for elucidating structural moieties in kerogen.

Nitric acid has also been used for the oxidation of kerogen (Robinson et al., 1963). However, it must be recognized that nitric acid reacts in different ways depending on the temperature, time, and concentration. Thus, in addition to the anticipated oxidation reactions, aromatic structures in the kerogen or even in the products are nitrated. However, nitric acid has been successfully used for investigating aliphatic structural units, and the data are often complementary to other structural studies.

Oxidants such as ozone (Rogers, 1973), air (Robinson et al., 1965), oxygen (Robinson et al., 1963), and hydrogen peroxide (Kinney and Leonard, 1961) have also been used for the oxidative degradation of kerogen. As in other cases, it is strongly recommended that structural data not be compiled in an absolute manner on the basis of one oxidant. The data from the various methods should be employed in a complementary manner so that an overall model can be compiled that explains the behavior of the kerogen under different reaction conditions.

3.4.4 Thermal Methods

Under favorable conditions, oxidation methods can give reasonably high recoveries of organic material but structural alteration does occur and some features are obliterated. A particularly attractive approach to minimize the formation of intractable residues—and the obliteration of important structural features—has been to heat the kerogen at moderate temperatures for prolonged periods and then to extract the degraded kerogen (Hubbard and Robinson, 1950). Thus, it may be possible to achieve complete conversion of the kerogen and recover >90% w/w as lower molecular weight (liquid) products. However, product alteration is also a feature of thermal reactions, but there is the distinct possibility that the thermal fragments that were allowed to escape from the reaction zone (by virtue of their volatility) would preserve some of the original (skeletal) features of the kerogen. Overall, the mild heat-soak-extraction method is a complementary method to the oxidation studies for providing lower molecular weight products for structural studies.

The use of micropyrolysis coupled with gas chromatography-mass spectrometry (GC-MS) can also provide valuable information about the structural units in kerogen (Schmit-Collerus and Prien, 1974). In addition to the online micropyrolysis-GC-MS studies, larger samples of kerogen have been pyrolyzed to obtain products that were fractionated by chromatography (ion exchange, complexation with ferric chloride [$FeCl_3$], and silica gel) into compound classes, and then by gel permeation chromatography into fractions of increasing molecular weight. These samples can then be investigated by conventional mass spectrometric techniques as well as by other spectroscopic approaches.

3.4.5 Acid-Catalyzed Hydrogenolysis

The use of hydrogenolysis in the presence of stannous chloride ($SnCl_2$) to degrade kerogen (Hubbard and Fester, 1958) affords good yields of liquid products for characterization purposes. Under these conditions, the majority of the heteroatoms are generally removed: nitrogen as ammonia, oxygen as carbon dioxide or as water, and sulfur as hydrogen sulfide. From these data, it was concluded that the nitrogen, oxygen, and sulfur functionalities comprise internuclear links (as opposed to their existence in ring systems) in the kerogen structure. These conclusions may appear to be at variance with other data, but this may be a result of the severe degradation conditions employed in the experiments, and examination under milder conditions is warranted.

3.4.6 Structural Models

The need to gather the very large mass of information about kerogen structure into a compact form useful for guiding research and development has led to models for kerogen structure. These models are not intended to depict the molecular structure of kerogen, at least not in the sense that the double helix describes the structure of deoxyribonucleic acid (DNA) or even in the sense that synthetic polymers are described in terms of monomers joined to form chains that have well-defined structures. The kerogen models represent attempts, based on the available data, to depict a collection of skeletal fragments and functional groups as a three-dimensional network in the most reasonable manner possible.

However, it must be recognized that no single analytical technique provides sufficient information to construct a precise model of the macromolecular structure of kerogen. Thus, most workers now use a multidimensional (multiple-technique) approach, but as is quite often the case, the techniques used by different workers have different strengths and may emphasize different features of kerogen structure. Sample-to-sample variation and the use of different isolation techniques also complicate the issue.

Several models have been proposed for the structure of kerogen in which a multidimensional approach has been employed. Such approaches are extremely valuable, since they bring together the results of several analytical methods. Indeed, the success of such approaches in the deduction of structural types in kerogen is also paralleled by the use of a similar approach to the deduction of the structural types that occur in the asphaltene fraction (Speight, 2007). Such structural models are of interest here because they give an overall picture of the perception of the kerogen structure and the models may even allow predictions of properties and behavior. If the model does not match actual behavior and properties, it must be reworked since it has little value.

A model derived for kerogen was based on the chromic acid oxidative degradation already discussed (Simoneit and Burlingame, 1974, and references cited therein). Additional information was provided by studies of the bitumen, again primarily by mass spectrometry, and was incorporated into the structural model that included regions of undefined structure containing trapped organic compounds of unknown nature and bearing side chains linked to the main structure by nonhydrolyzable C-C and hydrolyzable ester linkages. Ester linkages were also believed to be present and the model also includes an alicyclic (naphthene) ring. To understand this model, it is important to recall that the oxidation products (acids and ketones)—upon which most of this structure is based—represented only a fraction of the total organic carbon.

Another model for kerogen was developed on the basis of the data derived from the stepwise alkaline permanganate oxidation of kerogen, which produced high yields of carboxylic acids (Djuricic et al., 1971). Based on the oxidation results, a cross-linked macromolecular network structure was proposed. The most striking feature of this model is the predominance of straight-chain groups in the backbone of the network. The network bears both linear and branched side chains; branching points are indicated in the model by open circles. This model accommodates many important experimental observations, including reversible swelling and gel-like *rubbery* behavior in the swollen state, but does not satisfactorily account for the aromatic carbons observed by carbon 13 magnetic resonance spectroscopy, or the nitrogen and sulfur contents determined by elemental analysis.

Another kerogen model (Schmidt-Collerus and Prien, 1974) was assembled from the subunits identified by micropyrolysis-mass spectrometry studies. Key features of this model include formulation as a three-dimensional macromolecular network and a very uniform hydrocarbon portion composed mostly of small alicyclic and partially hydrogenated aromatic subunits with few heterocyclic rings. Long-chain alkylene and isoprenoid units and ethers serve as interconnecting bridges in this structure. Entrapped species (bitumen) include long-chain alkanes and both n-alkyl and branched-chain carboxylic acids. This model provides a useful view of the types and role of hydrocarbon units but de-emphasizes heteroatom functional groups and rings, presumably because groups containing these elements would not be detected efficiently by the micropyrolysis technique.

The structure of kerogen has also been probed by a wide variety of techniques, including stepwise alkaline permanganate and dichromate-acetic acid oxidation, electrochemical oxidation and reduction (in nonaqueous ethylenediamine-lithium chloride), and X-ray diffraction techniques

(Barakat and Yen, 1988, and references cited therein). It was concluded for that particular sample of kerogen that:

1. Aromaticity was low but isolated carbon-carbon double bonds were possible.
2. The structure was largely composed of three-to-four-ring naphthenes.
3. Oxygen was present mostly as esters and ethers.
4. The kerogen structure comprises a three-dimensional network and ethers serve as cross-links in this network.
5. Additional linkages are provided by disulfides, nitrogen heterocyclic groups, unsaturated isoprenoid chains, hydrogen bonding, and charge-transfer interactions.

Using these components as building blocks, a *multipolymer* network was envisaged. It was also pointed out that the extractable bitumen molecules could reside, more or less freely depending on their size, within the network.

To further account for the observed variations in the products obtained from the individual steps of stepwise permanganate oxidation, it was suggested that a *core plus shell* arrangement existed for the individual kerogen particles. The core was visualized as a cross-linked region containing most of the alkyl and alkylene chains and the bulk of the kerogen as naphthenic ring structures. On the other hand, the shell is more tightly cross-linked and contains most of the heteroatom functional groups and heterocyclic rings. This is interesting from the geochemical viewpoint, since the outer shell of this model is that part of the kerogen in contact with the mineral matrix; heteroatom functions tend to interact more strongly with minerals than do the hydrocarbon chains. Organic-mineral interactions in the resulting composite would then be ideally situated to hinder the physical separation of minerals from kerogen. This picture is consistent with the data of other workers (Siskin et al., 1987a, 1987b) on chemically assisted oil shale enrichment.

Another hypothetical model for kerogen is based on the results of a multidimensional approach to probing kerogen structure, which also included a detailed analysis of the functional groups in the kerogen (Scouten et al., 1987; Siskin et al., 1995). A comparison with other kerogen models serves to illustrate some of the key features of this model for kerogen. Aliphatic material is the most obvious feature of this model, and the aliphatic moieties are longer and more linear than those in other kerogen models. In addition, the aliphatic moieties are present both as alkylene bridges and as alkyl side chains, and secondary structures formed by paraffin-paraffin interactions are important in the kerogen. Naphthenic and partially hydrogenated aromatic rings also make an important contribution to the aliphatic moieties. The average ring system is only slightly larger than that in other kerogen models, but the size distribution is appreciably broader; a significant number of the larger four- to five-ring systems are present.

Yet another approach to deriving models for kerogen structure involves a more generalized procedure in which models representative of the three types of kerogen and of the asphaltene constituents from the corresponding oils as a function of maturity were developed (Behar and Vandenbroucke, 1987; Tissot and Espitalie, 1975). Emphasis in this work was placed on elucidating the chemistry of maturation for the three kerogen types and representing kerogen at the beginning of diagenesis (excluding the early stages of diagenesis, which is probably dominated by microbial action) (Behar and Vandenbroucke, 1987). These models provide an interesting view of the structural relationships between the three types of kerogen (Vandenbroucke, 2003).

REFERENCES

Anders, D.E., Doolittle, F.C., Robinson, W.E., 1975. Polar constituents isolated from Green River oil shale. Geochim. Cosmochim. Acta 39, 1423–1430.

Anders, D.E., Robinson, W.E., 1971. Cycloalkane constituents of the bitumens from Green River shale. Geochim. Cosmochim. Acta 35, 661.

Anderson, Y.C., Gardner, P.M., Whitehead, E.V., Anders, D.E., Robinson, W.E., 1969. The isolation of Steranes from Green River oil shale. Geochim. Cosmochim. Acta 33, 1304–1307.

Barakat, A.O., Yen, T.F., 1988. Novel identification of 17-beta (H)-hopanoids in Green River oil shale Kerogen. Energy Fuels 2, 105–108.

Behar, F., Vandenbroucke, M., 1987. Chemical modeling of kerogen. Org. Geochem. 11, 15–24.

Cummins, J.J., Robinson, W.E., 1964. Normal and isoprenoid hydrocarbons isolated from oil-shale bitumen. J. Chem. Eng. Data 9, 304–306.

Durand, B., 1980. Kerogen: Insoluble Organic Matter from Sedimentary Rocks. Editions Technip, Paris, France.

Erdman, J.G., 1981. Some chemical aspects of petroleum genesis. In: Atkinson, G., Zuckerman, J.J. (Eds.), Origin and Chemistry of Petroleum, Pergamon Press, New York.

Fenton, M.D., Henning, H., Ryden, R.L., 1981. In: Stauffer, H.C. (Ed.), Oil Shale, Tar Sands and Related Materials, American Chemical Society, Washington, DC, p. 315.

Fester. J.I., Robinson, W.E., 1966. Oxygen functional groups in Green River oil-shale kerogen and trona acids. Coal Science. Advances in Chemistry Series No. 55. American Chemical Society, Washington, DC, p. 22.

Forsman, J.P., 1963. Geochemistry of kerogen. In: Breger, I.A. (Ed.), Organic Geochemistry, Pergamon Press, Oxford, England, pp. 148–182.

Forsman, J.P., Hunt, J.M., 1958. Insoluble organic matter (kerogen) in sedimentary rocks of marine origin. In: Weeks, L.G. (Ed.), Habitat of Oil, American Association of Petroleum Geologists, Tulsa, OK, p. 747.

Himus, G., Basak, G.C., 1949. Analysis of coals and carbonaceous materials containing high percentages of inherent mineral matter. Fuel 28, 57–65.

Holmes, S.A., Thompson, L.F., 1981. Nitrogen-type distribution in hydrotreated shale oils: correlation with upgrading process conditions. In: Gary, J.H. (Ed.), Proceedings of the 14th Oil Shale Symposium, Colorado School of Mines Press, Golden, CO, p. 235.

Hubbard, A.S., Fester, J.I., 1958. Hydrogenolysis of Colorado oil-shale Kerogen. Ind. Eng. Chem. 3, 147–152.

Hubbard. A.B., Robinson, W.E., 1950. A Thermal Decomposition Study of Colorado Oil Shale. Report of Investigations No. 4744. United States Bureau of Mines, US Department of the Interior, Washington, DC.

Hubbard, A.B., Smith, H.N., Heady, H.H., Robinson, W.E., 1952. Method of Concentrating Kerogen in Colorado Oil Shale. Report of Investigations No. 5725. United States Bureau of Mines, US Department of the Interior, Washington, DC.

Kinney, C.R., Leonard, J.T., 1961. Ozonization of Chattanooga uraniferous black shale. J. Chem. Eng. Data 6, 474–476.

Lee, D.G., 1980. The Oxidation of Organic Compounds by Permanganate Ion and Hexavalent Chromium. Open Count Publishing Company, La Salle, IL.

McCollum, J.D., Wolff, W.F., 1990. Chemical beneficiation of shale kerogen. Energy Fuels 4, 11–14.

Regtop, R.A., Crisp, P.T., Ellis, J., 1982. Chemical characterization of shale oil from Rundle, Queensland. Fuel 61, 185–192.

Robinson, W.E., 1969. Kerogen of the Green River formation. In: Eglinton, G., Murphy, M.T.J. (Eds.), Organic Geochemistry, Springer-Verlag, Berlin, Germany, p. 181.

Robinson, W.E., Cummins, J.J., Dineen, G.U., 1965. Changes in Green River oil-shale paraffins with depth. Geochim. Cosmochim. Acta 29, 249.

Robinson, W.E., Dineen, G.U., 1967. Constitutional aspects of oil shale kerogen. In: Proceedings of the 7th World Petroleum Congress, Elsevier, Amsterdam, p. 669.

Robinson, W.E., Lawlor, D.L., Cummins, J.J., Fester, J.I., 1963. Oxidation of Colorado Oil Shale. Report of Investigations No. 6166. United States Bureau of Mines, US Department of the Interior, Washington, DC.

Rogers, M.P., 1973. Bibliography of Oil Shale and Shale Oil. Bureau of Mines Publications. Laramie Energy Research Center, United States Bureau of Mines, Laramie, WY.

Saxby, J.D., 1976. Chemical separation and characterization of kerogen from oil shale. In: Yen, T.F., Chilingarian, G.V. (Eds.), Oil Shale, Elsevier, Amsterdam, Netherlands, p. 103.

Schmidt-Collerus, J.J., Prien, C.H., 1974. Hydrocarbon structure of kerogen from oil shale of the Green River formation. Preprints. Div. Fuel Chem. Am. Chem. Soc. 19 (2), 100.

Scouten, C., 1990. Oil shale. In: Speight, J.G. (Ed.), Fuel Science and Technology Handbook, Marcel Dekker Inc., New York.

Scouten, C.G., Siskin, M., Rose, K.D., Aczel, T., Colgrove, S.G., Pabst, R.E., 1987. Detailed structural characterization of the organic material in Rundle Ramsay crossing oil shale. In: Proceedings of the 4th Australian Workshop on Oil Shale, Brisbane, Australia, pp. 94–100.

Simoneit, B.R.T., Burlingame, A.L., 1974. Study of organic matter in DSDP (JOIDES) cores, legs 10–15. In: Tissot, B., Bienner, F. (Eds.), Advances in Organic Geochemistry 1973, Editions Technip, Paris, France, p. 191.

Siskin, M., Brons, G., Payack, J.F., 1987a. Disruption of kerogen-mineral interactions in oil shales. Preprints. Div. Petrol. Chem. Am. Chem. Soc. 32 (1), 75.

Siskin, M., Brons, G., Payack, J.F., 1987b. Disruption of kerogen-mineral interactions in oil shales. Energy Fuels 1, 248–252.

Siskin, M., Scouten, C.G., Rose, K.D., Aczel, T., Colgrove, S.G., Pabst, Jr., R.E., 1995. Detailed structural characterization of the organic material in Rundle Ramsay crossing and Green River oil shales. In: Snape, C. (Ed.), Composition, Geochemistry and Conversion of Oil Shales, Kluwer Academic Publishers, Dordrecht, Netherlands, pp. 143–158.

Smith, J.W., Higby, L.W., 1960. Preparation of organic concentrate from Green River oil shale. Anal. Chem. 32, 1718–1719.

Speight, J.G., 2007. The Chemistry and Technology of Petroleum, fourth ed. CRC-Taylor and Francis Group, Boca Raton, FL.

Speight, J.G., 2008. Synthetic Fuels Handbook: Properties, Processes, and Performance. McGraw-Hill, New York.

Speight, J.G., 2009. Enhanced Recovery Methods for Heavy Oil and Tar Sands. Gulf Publishing Company, Houston, TX.

Speight, J.G., 2013. The Chemistry and Technology of Coal, third ed. CRC-Taylor and Francis Group, Boca Raton, FL.

Tissot, B., Deroo, G., Hood, A., 1978. Geochemical study of the Uinta basin: formation of petroleum from the Green River formation. Geochim. Cosmochim. Acta 42, 1469.

Tissot, B., Espitalie, J., 1975. L'Evolution Thermique de la Matiere Organiques des Sediments: Application d'une Simulation Mathematique. Revue Institut Français du Pétrole 30, 743.

Tissot, B., Welte, D.H., 1978. Petroleum Formation and Occurrence. Springer-Verlag, New York.

Uden, P.C., Siggia, S., Jensen, H.B. (Eds.), 1978. Analytical chemistry of liquid fuel sources. Advances in Chemistry Series No. 170. American Chemical Society, Washington, DC.

USGS, 1995. United States Geological Survey. Dictionary of Mining and Mineral-Related Terms. Second ed. Bureau of Mines & American Geological Institute. Special Publication SP 96-1, US Bureau of Mines, US Department of the Interior, Washington, DC.

Vadovic, C.J., 1983. Characterization of shales using sink float procedures. In: Miknis, F.P., McKay, J.F. (Eds.), Geochemistry and Chemistry of Oil Shales, Symposium Series No. 230. American Chemical Society, Washington, DC, p. 385.

Vandenbroucke, M., 2003. Kerogen: from types to models of chemical structure. Oil & Gas Science and Technology, Revue Institut Français du Pétrole 58, 243–269.

Vitorovic, D., 1980. Structure elucidation of kerogen by chemical methods. In: Durand, B (Ed.), Kerogen, Editions Technip, Paris, France, p. 301.

Whelan, J.K., Farrington, J.W. (Eds.), 1992. Organic Matter: Productivity, Accumulation, and Preservation in Recent and Ancient Sediments, Columbia University Press, New York.

Williams, P.F.V., Douglas, A.G., 1981. Kimmeridge oil shale: a study of organic maturation. In: Brooks, J. (Ed.), Organic Maturation Studies and Petroleum Exploration, Academic Press, London, England, p. 255.

CHAPTER 4

Mining and Retorting

4.1 INTRODUCTION

Shale oil is produced from oil shale by the thermal decomposition of its kerogen component. Oil shale must be heated to temperatures between 400 and 500°C (750–930°F). This heating process is necessary to convert the embedded sediments to kerogen oil and combustible gases. Generally, with solid fossil fuels, the yield of the volatile products depends mainly on the hydrogen content in the convertible solid fuel. Thus, compared with coal, oil-shale kerogen contains more hydrogen and can produce more oil and gas when thermally decomposed (Speight, 2007, 2008, 2013). From the standpoint of shale oil as a substitute for petroleum products, its composition is of great importance.

The thermal processing of oil shale to oil has quite a long history, and various facilities and technologies have been used. In principle, there are two methods for thermal processing: (1) low-temperature processing—semicoking or retorting—by heating the oil shale up to about 500°C (930°F), and (2) high-temperature processing—coking—heating up to 1000–1200°C (1830–2190°F).

A high-yield deposit of oil shale will yield 25 gallons of oil per ton of oil shale. Approximately 8 million tons of ore would need to be mined daily to meet one-quarter of the U.S. demand of 17–20 million barrels of oil per day, resulting in large quantities of spent shale that would need to be handled in an environmentally acceptable manner.

Production processes for the thermal treatment of oil shale deposits to produce shale oil fall into two categories as oil sand production processes: (1) ex situ production, which involves surface mining and processing and (2) in situ production methods, which involve heating the shale in place (underground) (Scouten, 1990; Yen, 1976).

In ex situ production, oil shale is mined, crushed, and then subjected to thermal processing at the surface in an oil shale retort. Both pyrolysis and combustion have been used to treat oil shale in a surface retort. In in situ production, the shale is left in place and the retorting (e.g., heating) of the shale occurs in the ground.

Generally, surface processing consists of three major steps: (1) oil shale mining and ore preparation, (2) pyrolysis of oil shale to produce kerogen oil, and (3) processing kerogen oil to produce refinery feedstock and high-value chemicals. For deeper, thicker deposits, not as amenable to surface or deep-mining methods, shale oil can be produced by in situ technology. In situ processes minimize, or in the case of true in situ eliminate, the need for mining and surface pyrolysis, by heating the resource in its natural depositional setting.

Depending on the depth and other characteristics of the target oil shale deposits, either surface mining or underground mining methods may be used. According to the method of heating, each method, in turn, can be further categorized (Table 2.1) (Burnham and McConaghy, 2006). Another way in which the various retorting processes differ is the manner by which heat is provided to the shale by hot gas: (1) by a solid heat carrier or (2) by conduction through a heated wall.

After mining, the oil shale is transported to a facility for retorting, after which the oil must be upgraded by further processing before it can be sent to a refinery, and the spent shale must be disposed of, often by putting it back into the mine. Eventually, the mined land is reclaimed. Both mining and processing of the oil shale involve a variety of environmental impacts, such as global warming and greenhouse gas emissions, disturbance of the mined land, disposal of the spent shale, use of water resources, and impacts on quality of air and water.

It is noteworthy at this point that any development of western oil shale resources will require water for plant operations, supporting infrastructure, and the associated economic growth in the region. Although some new oil shale technologies are claimed to significantly reduce process water requirements, stable and secure sources of significant volumes of water may still be required for commercial-scale oil shale development. The largest demands for water are expected to be for land reclamation and to support the population and economic growth associated with the oil shale activity.

4.2 MINING

For aboveground retorting, oil shale is to be removed from the deposit by mining. The shale excavated from mine varies greatly in size, from several millimeters to hundreds of millimeters, and even larger than 1000 mm. Pretreatment by crushing and screening is necessary to meet the demands of the retorting operation—commercial retorts have strict limits on the size range of the oil shale charge. Usually, the shale fraction can be divided in to lump shale and particulate shale as the feed for different types of retorts.

For lump shale, an internal hot gas carrier is usually used for supplying heat, whereas for particulate oil shale (less than 10 mm), an internal hot solid carrier is usually used. Combustion gas, pyrolysis gas, or retorted shale char can serve as the heat source(s). However, due to the low heat conductivity coefficient of lump oil shale, retorting takes longer because of the slow heating rate (only a few degrees per minute) and it can take several hours to reach the desired temperature. For particulate oil shale, due to its smaller size, the heating rate is higher, and the time required for retorting the oil shale is much shorter, only about several minutes or little more than 10 minutes.

In general, the developers of deposits in the United States are likely to use surface mining for the oil shale zones that are near the surface or that are situated with an overburden-to-pay ratio of less than about 1:1. Economic optimization methods can be used to select stripping ratios, optimum intercept, and cutoff grades.

Oil shale exhibits distinct bedding planes. These bedding planes can be used to an advantage during mining and crushing operations. Shear strength along the bedding planes is considerably less than across the planes, thereby, reducing operational demands. Thin overburden, attractive for surface mining, tends to be found along part of the margins of the southern Uinta Basin and the northern Piceance Creek Basin (Cashion, 1967).

The choice of how deep or selective to mine is an economic optimization issue. Numerous opportunities exist for the surface mining of ore averaging more than 25 gallons per ton, with overburden-to-pay ratios of less than 1, especially in Utah. In general, room and pillar mining is likely to be used for resources that outcrop along steep erosions. Horizontal adit, room and pillar mining was used successfully by UNOCAL.

In a mining-surface retorting process (*ex situ process*), oil shale rock is crushed and then conveyed to a retort. At the temperature in the retort (500–550°C, 930–1020°F), the organic constituents of the oil shale are converted to lower molecular weight distillable shale oil, which can also be a source of chemical products.

Open-pit mining has been the preferred method whenever the depth of the target resource is favorable to access through overburden removal. In general, open-pit mining is viable for resources where the overburden is less than 150 feet in thickness and where the ratio of overburden thickness to deposit thickness ratio is less than 1:1. Removing the ore may require blasting if the resource rock is consolidated, but in some cases exposed shale seams can be mined using a bulldozer. The physical properties of the oil shale, volume of operations, and project economics determine the choice of method and operation.

When the depth of the overburden is too great for economic surface mining, underground mining processes are required, which will necessitate a vertical, horizontal or directional access to the kerogen-bearing formation. Consequently, a strong *roof formation* must exist to prevent collapse or cave-ins, ventilation must be provided, and emergency egress must also be planned.

Room and pillar mining, as practiced in coal mining operations (Speight, 2013), has been the preferred underground mining option in the Green River Formation. Technology currently allows for cuts up to 90 feet in height to be made in the Green River Formation, where ore-bearing zones can be hundreds of feet thick. Mechanical *continuous miners* have also been selectively tested in this environment, with some degree of success.

In Colorado, suitable locations are at the north end and along the southern flank of the Piceance Creek Basin, where zones with a thickness of at least 25 feet and with yields of 35 gallons of shale oil per ton of oil shale exist throughout the area. In Utah, opportunities for 35 gallons of shale oil per ton of oil shale ore exist along Hell's Hole canyon, the White River, and Evacuation Creek. Because the pay zone is more than 1500 feet thick in some places, it is conceivable that open-pit mining could be applied even with 1000 feet of overburden.

It is worthy of note here that in recent years, Shell has experimented with a novel in situ process that shows promise for recovering oil from rich, thick resources lying beneath several hundreds to 1000 feet of overburden (Chapter 5). There are locations that could yield 1 million barrels of shale oil per acre of oil shale deposit and require, with minimum surface disturbance, fewer than 23 square miles to produce as much as 15 billion barrels (15×10^9 bbls) of shale oil over a 40-year lifetime of the project.

It also deserves mention that in the northern Piceance Creek Basin, zones of high-grade oil shale also contain rich concentrations of nahcolite (a mineral composed of sodium bicarbonate, $NaHCO_3$, also called thermokalite) and dawsonite [a mineral composed of sodium aluminum carbonate hydroxide, $NaAlCO_3(OH)_2$], which are high-value minerals that could be recovered through solution mining.

Advances in mining technology continue in other mineral exploitation industries, including the coal industry. Open-pit mining is a well-established technology in coal mining, tar sand mining, and hard-rock mining. Furthermore, room and pillar and underground mining have previously been proven at commercial scale for oil shale in the western United States. Costs for room and pillar mining will be higher than that for surface mining, but these costs may be partially offset by allowing access to richer ores. Indeed, current mining advances continue to reduce mining costs, lowering the cost of shale delivered to conventional retort facilities. Restoration approaches for depleted open-pit mines have been demonstrated, in both oil shale operations and other mining industries.

The advantages of ex situ processes include the following: (1) efficiency for organic matter recovery is high (approximately 70–90% w/w of the total organic content of the shale; (2) process operating variables are controlled; (3) undesirable process conditions can be minimized; (4) product recovery is relatively simple; and (5) process units can be used repeatedly for a large number of retorting operations.

However, there are disadvantages: (1) high operating cost because of the need to mine, crush, transport, and heat the oil shale; (2) because of the costs, the process is somewhat limited to rich-shale resources accessible for surface mining; (3) spent shale disposal; (4) the potential for underground water contamination; (5) the costs of revegetation of the site; (6) high capital investment for large-scale units; and (7) once the mine is depleted, some of the investment may have to be forsaken.

The liberated compounds from the oil shale retorting include gas and shale oil, which is collected, condensed, and upgraded into a liquid product that is considered, by some, to be equivalent to crude oil although this is not really the case (Chapter 6). This oil can be transported by a pipeline or a tanker to a refinery, where it is refined into the final product.

4.3 DIRECT RETORTING

Surface retorting involves (1) transporting the mined oil shale to the retort facility, (2) crushing the mined shale, (3) retorting, (4) recovering the raw shale oil, (5) upgrading the raw oil to marketable products, (6) disposing of the *spent* shale, and (7) reclamation of the mined land (Bartis et al., 2005). Retorting processes require mining more than a ton of shale to produce one barrel of oil. The mined shale is crushed to provide a desirable particle size and injected into a heated reactor (*retort*), where the temperature is increased to about 450°C (850°F). At this temperature, the kerogen decomposes to a mixture of liquid and gaseous products.

The United States, Russia, Estonia, Brazil, and China have developed several oil shale retorting technologies. In the United States, some companies have paid considerable attention to development work, and the Union Oil Company of California developed the rock-type (lump) retorting process with an oil shale processing capacity on the order of 10 000 tons—the highest capacity in the world (Barnet, 1982). Another company—TOSCO—developed moving-bed particulate retorting. However, all the retorts developed in the United States have not seen long-time continuous service in commercial production.

Numerous approaches to oil shale pyrolysis have been tested at pilot and semi-commercial scales during the 1980s (Scouten, 1990; Speight, 2008). The principal objectives of any retorting process are high yields, high energy efficiency, low residence time, and reliability. Larger-than-pilot-scale tests were made by TOSCO, Paraho, and Exxon. UNOCAL operated a full-scale commercial module. Occidental ran a large-scale modified in situ (MIS) project.

Retorting conditions for oil shale have a significant effect on the properties of the shale oil and the oil yield. Among them, heating the individual shale pieces is the prime consideration in developing retorting concepts and operating oil shale retorts. Therefore, the method of heat transfer to the raw shale provides a convenient way to classify the retorts. In a very broad sense, two different retorts can be distinguished: (1) directly and indirectly gas-heated retort and (2) directly solid-heated retort.

In direct heat retorting, some of the oil shale, char-bearing spent shale from previous retorting cycles, or some other fuel is combusted to provide heat for pyrolysis of the remaining oil shale, with the flame impinging directly on the oil shale undergoing retorting. Indirect heating, the more widely practiced alternative, involves the use of gases or solids that have been heated externally using a separate imported fuel or energy source and then introduced into the retort to exchange heat with the oil shale. Indirect heat sources include hot combustion gases or ashes from the combustion of an external fuel, ceramic balls that have been heated by an indirect source, or even the latent heat contained in retort ash from previous retort cycles. The flammable hydrocarbon gases and hydrogen produced during retorting are also sometimes burned to support the heating process.

In the directly solid-heated retort, heat is transferred by mixing hot solid heat carriers with fresh shale. This method involves a more complex heat carrier circulation system but has the advantages of high oil yield, easy scale-up, undiluted product gas, and direct use of spent shale. The Lurgi-Ruhrgas retort, the TOSCO retort, the Taciuk retort, and the Galoter retort are typical examples of directly solid-heated retorts.

In the directly gas-heated retort, heat is transferred by passing hot gases directly through the shale, mostly in a vertical shaft kiln. This kind of retort can be subdivided into the following two modes: (1) internal combustion mode and (2) external combustion mode. In the internal combustion mode retort, hot gases are generated by either combustion of residual carbon in the spent shale within the retort or combustion of some retort gases. However, the main disadvantage of a direct retort is that recovery efficiencies (80–90%) are lower compared with indirect retorting (US OTA, 1980).

Almost all of the commercial retorts and the retorts in development are internal heating retorts, that is, direct heating retorts. The external

heating retorts, that is, indirect heating retort, are such that heat is supplied from a hot medium to oil shale through a wall—this retorting
method is less popular because of the small unit capacity, expensive heat
transfer, and low thermal efficiency.

At present, the mature commercial technologies are as follows:
(1) Kiviter lump shale retorting—Estonia, (2) Galoter particulate oil shale
retorting—Estonia, (3) Petrosix lump shale retorting—Brazil, (4) the
Fushun retorting system—China, and (5) the scaled-up Taciuk particulate
shale retort, called AOSTRA Taciuk Processing (ATP)—Australia.

Initial attempts at oil shale pyrolysis were conducted in aboveground
retorts, using designs and technical approaches that were adapted from
technologies developed for other types of mineral resource recoveries.
There are numerous configurations for aboveground retorts, and these
are differentiated by the manner in which they produce the heat
energy needed for pyrolysis, how they deliver that heat energy to
the oil shale, the manner and extent to which the excess heat energy
is captured and recycled, and the manner and extent to which the
initial products of kerogen pyrolysis are used to augment subsequent
pyrolysis. Technologies include both direct and indirect heating of the
oil shale.

Although all retorts produce crude shale oil liquids, hydrocarbon
gases, and char, some have been designed to further treat these hydrocarbon fractions to produce syncrude. Other retorting processes contain
auxiliary features to treat problematic by-products such as nitrogen-
and sulfur-containing compounds; in some cases, they even convert
these compounds to saleable by-products.

In this section, various surface retorting processes are described. In
order not to show any preference, the retorts are listed alphabetically.

4.3.1 Alberta Taciuk Processor

The Alberta (AOSTRA) Taciuk Processor (ATP) was originally designed
to extract bitumen from tar sand (oil sand) (Speight, 2007, 2011) but has
found application in oil shale processing (Koszarycz et al., 1991; Schmidt,
2002, 2003; Taciuk, 2002; Taciuk and Turner, 1988). This retorting technology has been tested in Australia to process oil shale deposits found
in Central Queensland but has been superseded by the Paraho retort
(Schmidt, 2003).

In the unit, the rotary refractory is a high-melting-point material that lines furnaces. In a well-integrated process, about 30% of the energy from raw shale is sufficient to support the process energy requirements. With the Taciuk technology, about 20% of the energy from the raw shale is sufficient to support the process energy requirements. The rotary kiln retort combines direct and indirect heat transfer through recirculation of gas and hot solids. Some of the processed shale is mixed with the fresh feed to provide the energy, through solid-to-solid heat transfer, for combustion and retorting. This technology improves on previously explored retorting methods by increasing oil and gas yields, improving thermal efficiency, reducing process water use, and minimizing the residual coke on the spent shale. The system has been designed to reduce both gaseous and particulate emissions and to make disposal of the spent shale straightforward and efficient.

The Taciuk technology has not been tested and demonstrated on western U.S. oil shale reserves, and there is uncertainty regarding the use of this technology domestically due to the different composition of Colorado shale relative to the Australian shale (Johnson et al., 2004). Another potential difficulty in applying the Taciuk technology is that Colorado oil shale will generate more fine particles than Australian oil shale does (Andrews, 2006). However, other researchers have concluded that due to their richness, Colorado oil shales will be easier to process using the Taciuk technology with some process modifications (Berkovich et al., 2000).

The Oil Shale Exploration Company (OSEC) planned to apply the Taciuk technology to processing oil shale from the Green River Formation in Utah. The company was awarded a 160-acre (0.65 square kilometer) research, development, and demonstration lease by the Bureau of Land Management (BLM) in December 2006 for the White River Mine site in Uintah County, Utah. The plans were to produce the shale oil from approximately 50 000 tons (45 000 metric tons) of previously mined oil shale available at the site. However, on June 9, 2008, OSEC announced that it had signed an agreement with Petrobras and Mitsui according to which Petrobras agreed to undertake a technical, economic, and environmental commercial feasibility study of the Petrosix shale oil technology for the oil shale owned or leased by OSEC in Utah. Few results of the study were disclosed.

In March 2011, it was announced that Eesti Energia (an Estonian Energy Company) (Chapter 2) would acquire all of the shares of OSEC.

On March 15, 2011, the transaction was approved by the Committee on Foreign Investment in the United States. After acquiring the OSEC shares, Eesti Energia announced that it would conduct a new commercial study using its Enefit process.

The Australian Stuart project implemented the Taciuk technology in a multistage strategy—originally, the technology was chosen because of (1) the simple design and energy self-sufficiency, (2) minimal process water requirements, (3) ability to handle fines, and (4) high shale oil yield (Johnson et al., 2004). In Stage 1, the production level was 4500 barrels of shale oil per day and produced 1.3 million barrels of shale between 1999 and 2004. By Stage 3, production levels were projected to be 200 000 barrels of shale oil per day. However, the Taciuk processor achieved only 55% capacity in a sustained trial due to mechanical problems and plugging by fine solids. The project was stopped in late 2004 for further evaluation, and the operation subsequently went out of business.

Queensland Energy Resources (Australia) assessed the possibilities for future commercial operation of the Stuart project, and spent the 2005–2007 period testing indigenous Australian oil shale at a pilot plant in Colorado. The results of the tests showed that, using the Paraho process, an oil-shale-to-shale-oil and liquid products business could be operated in Queensland.

4.3.2 Chattanooga Process

Central to the Chattanooga process is a pressurized fluid bed reactor and associated fired hydrogen heater. Conversion occurs in a relatively low temperature ($<535°C$, $1000°F$), noncombustion environment. With modifications only to its feed system, the reactor can convert oil-bearing materials such as oil sand, oil shale, and liquid bitumen via thermal cracking and hydrogenation into hydrocarbon vapors and spent solids.

Hydrogen is used as the heat conveyor to the reactor, reactor bed fluidizing gas, and reactant. Hydrogen is heated in an adjacent fired heater fueled by process off-gases and supplemental gas or product oil, depending on economic conditions. This flexibility minimizes or eliminates natural gas requirements.

For the heater and the associated hydrogen plant reformer, combustion air is preheated by cooling the spent sand or shale discharged from the reactor. Reactor overhead gases are cleaned of particulate solids in a hot gas filter, cooled and hydrocarbon products condensed, and separated from the gas stream. The liquid product produced at this stage may be lightly hydrotreated to produce a very low-sulfur, high-grade synthetic crude oil.

The excess hydrogen, low-boiling liquids, and acid gases are passed through an amine-scrubbing system to remove hydrogen sulfide, which is converted to elemental sulfur. The excess hydrogen and low-boiling liquids, now stripped of the acid gases, together with new makeup hydrogen are admitted to a turbine-driven centrifugal compressor for recompression and recycling. Steam for the turbine is generated by recovering the waste heat from the fired heater. Compressor power requirements are minimized by maintaining a low pressure drop around the process loop.

A slip stream of recycle gases is taken from the compressor discharge and passed through a purification system to remove hydrocarbon gases produced in the reactor. The purified hydrogen gas stream is returned to the compressor inlet. The hydrocarbon gases may be used as feedstock to the integrated hydrogen plant, thus, again minimizing the requirement for purchased natural gas.

Use of hydrogen in the initial phase of the process greatly enhances the quality of the product and reduces the need for extreme hydrotreating in downstream operations. Recovery of waste heat, power cogeneration, and the utilization of the hydrocarbon gases produced in the reactor as feedstock for the hydrogen plant make the Chattanooga process virtually self-sufficient by obtaining its energy requirements from the primary plant feedstock.

4.3.3 The EcoShale In-Capsule Process

The EcoShale In-Capsule process integrates surface mining with a relatively low-temperature roasting method that occurs in an impoundment that is constructed in the void space created by the shale mining excavation. A similar low-temperature concept was used in Germany during World War II (Kogerman, 1997).

When filled with shale, the capsule is heated using pipes circulating hot gases derived from burning natural gas or its own produced gases. To maximize energy efficiency, the process heat used in one capsule can be recovered by circulating lower temperature gases, which transfer the remaining heat into adjacent capsules. The lower-temperature slower roasting approach also minimizes carbon dioxide emissions and is amenable to carbon capture and sequestration. The unique impoundment approach allows for rapid reclamation and approximate restoration of the topography.

4.3.4 Fushun Generator Type Retorting

Similar to the Estonian and Russian Generator type retort, the Fushun type retort has been developed and used in commercial production for more than 70 years in Fushun, China (He, 2004; Hou, 1986; Zhao and He, 2005; Zhou, 1995). The retort is of vertical cylindrical type, with an outside being steel plate and the inside lined with fire bricks; its inner diameter is about 10 feet with a height of approximately 30 feet.

The oil shale (sized to 10–75 mm) is fed from the top of the retort, with the size of 10–75 mm; at the upper section (pyrolysis section) of the retort, the oil shale is dried and heated by the hot ascending gaseous heat carrier and pyrolyzed at about 500°C; the oil-gaseous vapor produced exits from the top of the retort, the oil shale is converted to shale coke, and goes to the lower part (gasification section) of the retort; it reacts with the ascending air-steam (coming from the bottom of the retort); it is gasified and combusted to the shale ash; the air-steam reacts with the coke to form hot gas and flows to the upper part of the retort to heat the oil shale. At the middle of the retort, a hot recirculating gas as the supplementary hot gas carrier is introduced to heat the oil shale, and this recirculating gas is part of the retort exit gas, after it is cooled in the condensation system (shale oil is condensed), and it is again heated in a recuperator to 500–700°C, then back to the retort. The shale ash exits from the water dish at the bottom of the retort.

Twenty Fushun retorts share one condensation system; that is, the exit gas evolved from 20 Fushun retorts flows together to a collecting tube, and then successively to a washing tower, gas blower, and cooling tower, where the shale oil is condensed; a part of the retort gas coming from the gas blower is introduced as fuel to the recuperator. Meanwhile a part of the retort gas is introduced to an another recuperator, where it is heated and recirculated to the middle of the retort as a hot gas carrier for heating the oil shale in the retort; the remaining retort gas coming from the cooling tower is coming out of the condensation system as surplus gas.

The characteristics of Fushun retort are as follows: the potential heat of the fixed carbon of the shale coke is partly used; thus, high thermal efficiency is obtained, but due to the addition of air into the retort, after combustion, the nitrogen dilutes the pyrolysis gas, which gives the retort-exit gas a low calorific value; furthermore, the excess oxygen coming to the upper part of the retort will burn out a part of the shale oil produced, thus reducing the shale oil yield greatly.

The oil yield of the Fushun retort accounts for about 65% of Fisher assay. The daily capacity of the retort is only 100–200 tons. The Fushun type retort is suitable for small oil shale retorting plant and for processing lean oil shale with low gas yield.

4.3.5 Enefit Process
The Enefit process is a modification of the Galoter process being developed by Enefit Outotec Technology. In this process, the Galoter technology is combined with proven circulating fluidized bed combustion technology used in coal-fired power plants and mineral processing. Oil shale particles and hot oil shale ash are mixed in a rotary drum as in the classical Galoter process. The primary modification is replacement of the Galoter semicoke furnace with a fluidized bed combustion furnace. The Enefit process also incorporates fluid bed ash cooler and waste heat boiler commonly used in coal-fired boilers to convert the waste heat to steam for power generation.

Compared to the traditional Galoter technology, the Enefit process allows complete combustion of carbonaceous residue, improved energy efficiency by maximum utilization of waste heat, and less water use for quenching. According to its promoters, the Enefit process has a lower retorting time compared to the classical Galoter process, and therefore, it has a greater throughput. Avoidance of moving parts in the retorting zones increases their durability.

4.3.6 Galoter Retort
The Galoter retort (the newest modifications of which are the Enefit process and the Petroter process) is a technology for the production of shale oil. In this process, oil shale is decomposed into shale oil, gas, and spent residue. The process was developed in the 1950s, and it is used commercially for shale oil production in Estonia. There are projects further developing this technology and expanding its usage, for example, in Jordan and the United States.

The retort is a near-horizontal slightly inclined cylindrical rotating one, and the feed oil shale is crushed and sized to approximately 25 mm. The shale ash is used as a solid heat carrier. In the horizontal cylindrical retort, the dried oil shale is mixed with the hot ash carrier, and is heated to 500°C. Then, it is pyrolyzed for about 20 min, and shale coke is formed and it moves with the ash from the retort into the vertical fluidized combustion chamber, where it is combusted with the incoming up-flowing air. Shale coke is converted into shale ash, having a temperature of the order of 700–800°C (1290–1470°F).

The shale ash is separated from the hot flue gas in the cyclone and is mixed with the dried oil shale. Both are introduced into the retort—the dried oil shale is heated and pyrolyzed, and the shale ash is recirculated with the shale coke. The hot flue gas leaving the cyclone is introduced to the waste heat boiler and then to the fluidized drier for drying the oil shale feed. The shale oil vapor exits the retort and is cooled successively; thus, heavy oil, light oil, naphtha fractions, and high-calorific gas are obtained.

Two Galoter solid heat carrier retorts, each with a processing capacity of 3000 tons of oil shale per day, were built at Narva Power Plant, Estonia (Golubev, 2003; Opik et al., 2001). Technological chemical efficiency accounts for 73–78% with an oil yield of 85–90% of Fisher assay. The retort gas contains 30% light olefins and may be used for producing petrochemicals or as town gas; besides, the oil shale is used as feed (Senchugov and Kaidalov, 1997).

This retorting technology is complex, requiring more equipment and machines; its operation is not easy. Estonia and Russia have spent much time and money to develop the technology – more than 50 years to take the project from laboratory to pilot plant to commercial scale.

4.3.7 Gas Combustion Retorting Process

The U.S. Bureau of Mines gas combustion retorting process uses a vertical, refractory-lined vessel (similar in operating concept to a moving-bed reactor used for coal gasification) through which the crushed shale moves downward by gravity, countercurrent to the retorting gases (Figure 4.1) (Matzick et al., 1966; US OTA, 1980). Recycled gases enter the bottom of the retort and are heated by the hot spent shale as they pass upward through the vessel. Air and some additional recycle gases are injected into the retort through a distributor system located above the heat

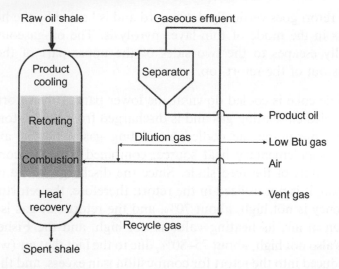

Fig. 4.1. Gas combustion retort.

recovery zone to mix with the rising hot recycled gases. Combustion of the gases and some residual carbon heats the shale immediately above the combustion zone to the retorting temperature. Oil vapors and gases are cooled by the incoming shale, and the oil leaves the top of the retort as a mist.

The benefits of this system include the following: (1) energy is recovered from the retorted shale giving the unit high thermal efficiency, and (2) an important consideration in arid regions (such as the western United States) is that there are no requirements for cooling water.

The U.S. Bureau of Mines developed and tested this retorting system during the 1980s specifically for the Green River Shale Formation. However, the project was terminated prior to the operation of the largest of three pilot plants (Dinneen, 1976; US OTA, 1980).

4.3.8 Kiviter Retort
The Estonia Kiviter retort (Sonne and Doilov, 2003; Yefimov and Doilov, 1999) is a kind of vertical cylindrical retort, having rectangular combustion chambers at the middle of the upper part of the retort and also at the two sides of the middle part of the retort. The combustion chambers are equipped with air and recirculating gas nozzles; the combustion takes place, and the hot combusted gas formed horizontally comes to the two pyrolysis chamber, where the oil shale fed from the

top of the retort goes vertically downward and is heated by the hot combusted gas in the mode of thin-layer pyrolysis. The oil-gaseous vapor
horizontally escapes to the two sides of the upper part of the retort
and comes out of the retort top.

The shale coke is cooled down at the lower part of the retort by the
upward cooled circulating gas and is discharged from the bottom water
seal; in the meantime the cooled circulating gas is heated and goes
upward as supplementary heat source, combined with the combusted
gas for pyrolysis of the feed shale. Since the discharged coke contains
fixed carbon, it is not utilized in the retort; therefore, the retorting thermal efficiency is not high, about 70%, and the retort exit gas is diluted
by nitrogen in air, its heating value is not high, and the Fisher assay
oil yield is also not high, about 75–80%, due to the fact that air (with oxygen) introduced into the retort for combustion is in excess, and the excess
oxygen burns out some shale oil produced, or mainly because part of the
shale oil produced is pyrolyzed by the hot combusted gas.

The daily capacity of the retort accounts for 1000 tons oil shale with
a size of 10–125 mm (mainly 25–100 mm), the electricity consumption
for processing 1 ton oil shale is about 14–18 kwh, steam (5–8 bars) 15–20 kg,
and water 0.2–0.5 m^3. Two Kiviter retorts, each with a daily capacity
of 1000 tons oil shale, have been well operated at Estonia Viru Keemia
Group (VKG), Kohtla Jarve. This retort is suitable for medium and small
shale oil plants.

4.3.9 Lurgi-Ruhrgas Process
The Lurgi-Ruhrgas technology was developed in Germany for the production of pipeline-quality gas through the devolatilization of coal fines
(Figure 4.2). The process was designed not only to retort kerogen but also
to refine the resulting hydrocarbons into saleable liquid fractions similar
to the products from petroleum. The technology has operated at commercial scales for the devolatilization of lignite fines, the production of
char fines for briquettes from sub-bituminous coal, and the cracking of
naphtha and crude oil to produce olefins (Speight, 2007, 2008, 2013).

In the process, the crushed and sized oil shale (<0.25 inch) is fed through
a feed hopper and mixed with as much as six to eight times the volume of
a mixture of the hot spent shale and sand with a nominal temperature
of 630°C (1165°F) and was conveyed up a lift pipe. This mixing raises
the average temperature of the raw shale to 530°C (985°F), a temperature

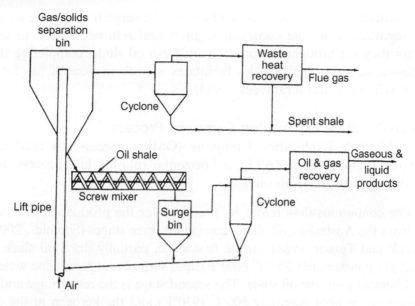

Fig. 4.2. Lurgi–Ruhrgas process.

sufficient to cause the evolution of gas, shale oil vapor, and water vapor. The solid mixture is then delivered to a surge hopper to await additional processing in which more residual oil components will be distilled off. The sand, introduced as a heat carrier, is recovered and recycled. The mixture is then returned to the bottom of the lift pipe and allowed to interact with hot combustion air at 400°C (750°F). The carbonaceous fraction is burned as the mixture is raised pneumatically up the lift pipe and transferred to a collection bin where the spent shale fines are separated from gases.

The hydrocarbon gases and oil vapors are processed through a series of scrubbers and coolers to eventually be recovered as condensable liquids and gases. Because the particle size of the oil shale feedstock is small, management of fines is critical throughout the process and involves the use of sedimentation and centrifuging as well as numerous cyclones and electrostatic precipitators.

Retorted shale from the mixer passes through the hopper to the bottom of the lift pipe, with the dust from the cyclone. Preheated air introduced at the bottom of the pipe carries the solids up to the surge bin. The solids are heated by the combustion of the residual char in the shale to approximately 550°C (1020°F). In the case when the residual char is not sufficient for this, fuel gas is added. In the surge bin, the hot solids are separated from the combustion gases and returned to the mixer, where they are brought in contact with fresh oil shale, completing the cycle—a surge bin with baffles facilitates a uniform flow of the feedstock (Kennedy and Krambeck, 1984).

4.3.10 Oil Shale Exploration Company Process
The Oil Shale Exploration Company (OSEC) process has used the Alberta Taciuk Process (ATP), a horizontal rotating kiln process, for the development of Utah oil shale.

The continuous-flow retort, as proposed for the production of shale oil from the Australian oil shale, comprised three stages (Schmidt, 2003; Taciuk and Turner, 1988). In the first stage, partially dried oil shale is dried at approximately 250°C (480°F), and surface and crystalline waters are liberated from the oil shale. The second stage is the retort stage and is carried out at approximately 500°C (930°F) and the kerogen in the oil shale is decomposed to produce the shale oil and hydrocarbon. The third and final stage involves the combustion of the retorted oil shale at 750°C (1380°F) and is coupled with hot solids recycle to heat the oil shale in earlier stages.

The retort has been used in the Stuart oil shale project, Australia, which produced more than 1.5 million barrels of shale oil (US DOE, 2007). However, the Stuart project has recently opted for the Paraho retort in place of the Taciuk retort.

4.3.11 Paraho Retort
The Paraho retort has been in service in oil shale fields in both Colorado and Brazil. Two versions exist: (1) the direct heating mode retort and (2) the indirect heating mode retort—both retorts use vertical retorting chambers.

In the direct mode retort (Figure 4.3), some of the raw shale is ignited in the combustion zone of the retort to produce the heat that pyrolyzes the remaining oil shale present in higher zones. In the indirect mode retort, heat is generated in a separate combustion chamber and delivered to the lowermost portion of the retorting chamber.

In the direct mode Paraho retort, crushed and sized oil shale is fed into the top of the vertical retorting vessel. At the same time, spent shale

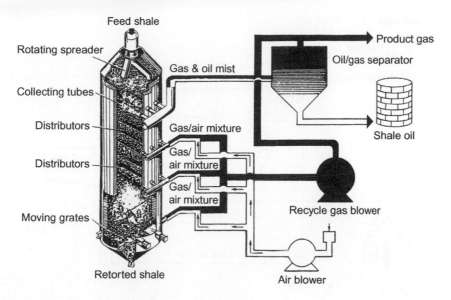

Fig. 4.3. Paraho retort—direct heating mode (Scouten, 1990).

(previously retorted oil shale that contains solid carbonaceous char) is ignited in a lower level of the retort. Hot combustion gases rise through the descending raw shale to decompose the kerogen. Shale oil vapor formed in the uppermost portion of the retort is removed for further upgrading in independent facilities. Any gases are cleaned for sale, while a small portion is returned to the retort and combusted together with the spent shale.

4.3.12 Petrosix Retort

The Brazilian Petrosix retort is also a vertical cylindrical retort (Figure 4.4) (Hohman, 1992; Martignoni, 2002; Scouten, 1990).

The Petrosix retort has an upper pyrolysis section and a lower shale coke cooling section. Shale oil is produced at 500°C (930°F) and is recovered at the top of the retort. The retort off-gas is cooled successively by cyclone, electric precipitator, and spraying tower for condensation, and part of the cooled retort gas is used as fuel in a tubular heater. Another part of the cooled retort gas is heated to a temperature in excess of 500°C (930°F), and recirculated back to the middle of the retort as a hot gas carrier for heating and pyrolyzing the oil shale feed. Another part of the cooled retort gas is circulated and enters into the bottom of the retort where it cools the hot shale. The gas now heated

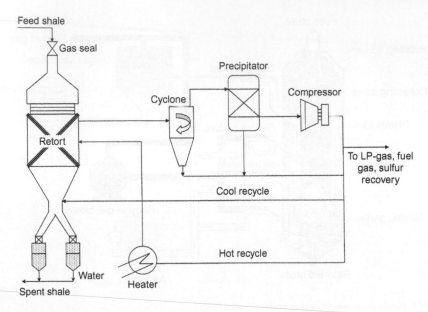

Fig. 4.4. Petrosix retort.

up again ascends into the pyrolysis section as supplementary heat source for heating the oil shale feed. The cooled shale is discharged from the water seals at the retort bottom.

The advantages of this method are as follows: (1) the retort capacity is high, (2) the off-gas has high calorific value—it is not diluted by nitrogen, and (3) the oil yield is high, reaching 85–90% of the Fischer assay. However, the thermal efficiency is variable depending on whether the carbon deposited on the shale is used.

Petrobras Company at Sao Mateus do Sul built two large Petrosix retorts, one with an inside diameter of 18 feet (built in 1981), for processing lump (6–50 mm) oil shale, the other with an inside diameter of 38 feet (built in 1991). The 38-foot diameter retort has a daily capacity of 1600 tons, with the capital cost of US$35 000 000. The 18-foot diameter retort has a daily capacity of 1600 tons, with the capital cost of US$35 000 000.

4.3.13 Petroter Retort

Petroter technology uses a modified Galoter retort for a pyrolysis (semi-coking) process to treat fine-grained oil shale (0–25 mm fraction) with a solid heat carrier. Through a mixture of oil shale and heated ash in the absence of air, the heating and (by sufficient temperature) decomposition of the oil shale organic matter occur to produce shale oil and gases.

The oil shale pyrolysis process is accomplished in a drum-rotating reactor in the absence of air, at a temperature of 450–500°C, (840 to 930°F) due to the mixture of oil shale and hot ash, which is used as a solid heat carrier. The vapor-gas product mixture that appears in the reactor during the process is fed through several process vessels for the removal of ash and any other mechanical impurities after which the liquid and gases are subjected to a distillation process to produce liquid products and gas with high calorific value.

The liquid products are fed to other units for loading as final products or for further processing. The gas is fed to the heat power plant for heat and power production. Steam is fed to the heat power plant for power production. The by-products of this process include phenol water, flue gases, and ash from thermal processing.

4.3.14 Superior Oil Circular Grate Retorting Process

This retort—a counterflow, gas-to-solid heat exchange process conducted in an enclosed circular grate, which has been used for processing various ores with a relatively high reliability factor—offers environmental advantages over other retorts.

In the process, oil shale (0.25–4.0 inch) is added, rotated to the first segment of the retort, and heated by a continuously circulating gas medium. Volatilized shale oil mixes with the circulating gas and, together with water, is periodically removed from the gas stream. The partially pyrolyzed shale rotates to the next segment of the retort where it is partially oxidized to complete the kerogen pyrolysis and oil evolution. The spent shale cools in the next segment of the grate as it yields heat to the circulating gas. Additional heat is added to the first segment of the grate where initial pyrolysis of raw shale takes place either through direct or indirect combustion of gases recovered from the previous shale retorting.

Temperature control is excellent, resulting in high hydrocarbon recovery rates and relatively minor amounts of sintering of the inorganic phase of the shale. Recovery yields of shale oil are on the order of 98% w/w of Fisher assay data.

From an environmental perspective, the circular grate is a sealed operation with hooded enclosures above the grate to capture hydrocarbon gases and shale oil vapor—water troughs (water seals) are installed below the grate where the spent shale is discharged. The water seals not only prevent the leakage of gas and vapor mist but also provide for the moistening the spent shale that is necessary for safe handling and disposal.

4.3.15 Superior Oil Multimineral Process

The superior oil multimineral process (also known as the McDowell–Wellman process or circular grate process) heats oil shale in a sealed horizontal segmented retort to produce shale oil, gas, and spent residue. The process is suitable for processing mineral-rich oil shales, such as that in the Piceance Basin, and has a relatively high reliability and high oil yield.

The multimineral process also produces minerals such as nahcolite alumina (Al_2O_3) and soda ash in addition to shale oil and gas (Matar, 1982). The process is basically a four-step operation for oil shale that contains recoverable concentrations of shale oil, nahcolite ($NaHCO_3$), and dawsonite [a sodium–aluminum salt, $Na_3Al(CO_3)_3 \cdot Al(OH)_3$] (Matar, 1982).

The process was originally developed as a stirred bed, low-Btu coal gasifier. The continuously fed, circular-moving-grate retort used in this process is a proven, reliable piece of hardware that provides accurate temperature control, separate process zones, and a water seal that eliminates environmental contamination.

The nahcolite has been tested as a dry scrubbing agent to absorb sulfurous and nitrous oxides. In the retort, the dawsonite in the shale is decomposed into alumina and soda ash. After the shale is leached with recycled liquor and makeup water from the saline subsurface aquifer, the liquid is seeded and the pH is lowered to recover the alumina. This alumina can be extracted and recovered at a competitive price with alumina from bauxite. The soda ash is recovered by evaporation and can be used for a variety of industrial applications such as neutralizing agents. The leached spent shale is then returned to the process.

4.3.16 Union B Retort

This retort was developed by the Union Oil Company of California (UNOCAL) and is an example of hot inert gas retorting. This process does not require cooling water.

In the process, crushed shale (0.13–2.00 inch) is fed through two chutes to a solid pump that moves the shale upward through the retort. The shale is heated to retorting temperatures by interaction with a counterflow of hot recycle gas (510–538°C, 950–1000°F), resulting in the evolution of oil shale vapor and gas. Heat is supplied by combustion of the organic matter remaining on the retorted oil shale and is transferred to the (raw) oil shale by direct gas-to-solid exchange. This mixture is forced downward by the flow of recycle gas and cooled by contact with the cold shale entering the retort in the lower section of the retort. Gas and condensed liquids are separated at the bottom of the retort. The liquids are removed and are further treated for removal of water, solids, and arsenic salts. The gases are sent to a preheater and returned to the retort for the recovery of heat energy by burning.

Pollution control devices are integrated into the design for the removal of hydrogen sulfide (H_2S) gas and ammonia (NH_3) gas produced during retorting and for the treatment of process waters recovered from oil-water separations. Treated waters are recycled, used for cooling the spent shale, or delivered to mining and handling operations and used to moisten the shale for fugitive dust control.

The reducing atmosphere maintained in the retort results in the removal of sulfur and nitrogen compounds through the formation of hydrogen sulfide and ammonia, respectively, both of which are subsequently captured. Forcing the hot newly formed oil vapors to contact the cooler shale as it enters the retort results in rapid quenching of the vapors. Additional treatment of the initially formed shale oil and the removal of heavy metals, such as arsenic, results in a final product recovered from the retort that can be used directly as a low-sulfur fuel or delivered to conventional refineries for additional refining.

A major issue was the formation of fine solids by decrepitation of the shale during retorting; the fine solids created problems in controlling solids flow in the retort and cooling shafts.

4.4 INDIRECT RETORTING

In the indirectly gas-heated retort, oil shale is heated through a barrier wall. This type of retort was used mainly before 1960 and is not being developed further because of (1) its small capacity, (2) expensive heat transfer, and (3) low thermal efficiency.

4.4.1 Paraho Retort

The Paraho Development Corporation developed new vertical shaft kiln hardware and process techniques and confirmed new technology in the 1960s by building three large commercial lime kilns. In the 1970s, the company adapted their lime kiln technology to oil shale retorting. Paraho obtained a lease from the U.S. Department of the Interior in May 1972 for the use of the U.S. Bureau of Mines oil shale facility at Anvil Points near Rifle, Colorado, to demonstrate their retorting technology.

In the *indirect mode* Paraho retort (Figure 4.5), the portion of the vertical retorting chamber that was used for oil shale combustion in the direct mode is now the region of the retort chamber into which externally heated fuel gas is introduced. No combustion occurs within the retorting chamber. The separate combustion process is typically fueled by commercial fuels (natural gas, diesel, propane, etc.) that are often augmented with a portion of the fuel gas recovered from the retorting operation.

In the process, finely ground oil shale enters a feed hopper on the top of the retort after which, in a continuous moving bed, the oil shale flows downward consecutively through the mist formation, retorting, combustion, and cooling zones. As the shale descends, heat is efficiently

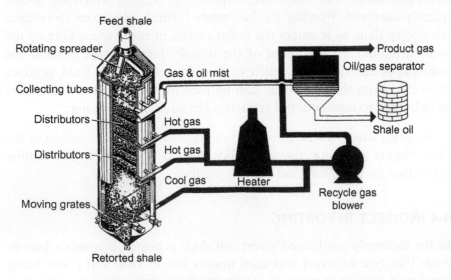

Fig. 4.5. Paraho retort—indirect heating mode (Scouten, 1990).

exchanged with a counter current flow of recycle gas, which is introduced into the retort at different levels by three specific-purpose gas-air and gas distributors. Near the top of the retort, the ambient temperature shale is warmed by rising hot oil vapors and gas, which, in turn, are cooled to form an oil mist that is entrained in the gas.

Although the direct and indirect mode Paraho retorts are very similar in operation, they offer sufficiently different operating conditions to change the composition of the recovered crude shale oils and gases. Oil vapors and mists leave the direct mode retort at approximately 60°C (140°F), whereas the vapors and gases in the indirect mode leave the retorting vessel at 135°C (280°F) and have as much as nine times higher heating values than gases and vapors recovered from the direct mode retort (102 Btu/ft^3 and 885 Btu/ft^3)—oil vapor and mists recovered from the direct mode are diluted with combustion gases from the combustion of the spent shale at the bottom portion of the retort.

The characteristics of the recovered raw shale oil are different for the direct and indirect mode retorts, but each has characteristics similar to shale oils recovered from other retorts using similar shale heating mechanisms (direct vs. indirect). In addition, gases from the indirect mode retorts have much lower levels of carbon dioxide but generally higher levels of hydrogen sulfide, ammonia, and hydrogen, which are believed to be the result of the indirect mode retort having much less of an oxidizing environment than the direct mode retort (EPA, 1979).

4.4.2 Pumpherston Retort

The Pumpherston retort emerged as the most successful retort of the Scottish oil shale industry in the second half of the nineteenth century. The plant was established in Scotland in 1847 and consisted of two benches with 52 retorts—each with a capacity of 10 tons per day and daily oil production of about 530 barrels. Acid was recovered from the acid tars and used to produce ammonium sulfate, and the tar was used as a refinery fuel.

The Pumpherston retort had the advantages of producing undiluted shale gas and had few restrictions on the grain size of the feed shale. The retort was closed down in 1963 because of competition from cheap and imported petroleum.

4.4.3 Salermo Retort

In the Salermo retort, small oil shale particles (<25 mm) were progressively shoveled through a series of 36 semicircular troughs (16 inch in diameter) heated from the underside by combustion of noncondensable gases. Each of the 10 externally heated retorts processed approximately 670–70 tons of oil shale per day to produce 810 barrels of shale oil per day, approximately 90% of Fischer assay. The spent shale containing large quantities of residual carbon (up to 40%) was discarded (Steele, 1979.).

This type of retort was operated near Ermelo, Transvaal (South Africa), in 1935. The plant was closed in about 1962; by that time, the plant had exhausted its oil shale resources. At present, one Salermo retort is still in operation in Clerbagnoix, France, to produce high-sulfur shale oil.

4.4.4 TOSCO II Process

The TOSCO (The Oil Shale Corporation) process used a rotating kiln that was reminiscent of a cement kiln in which heat was transferred to the shale by ceramic balls heated in an exterior burner (Figure 4.6) (Whitcombe and Vawter, 1976). The process, which was initiated in the 1960s and 1970s and developed by the Oil Shale Corporation, is more correctly described as a retorting/upgrading process (US OTA, 1980).

In the process, oil shale—crushed and sized (nominally to 112 inch) raw oil shale and preheated to 260°C (500°F) by interaction with flue

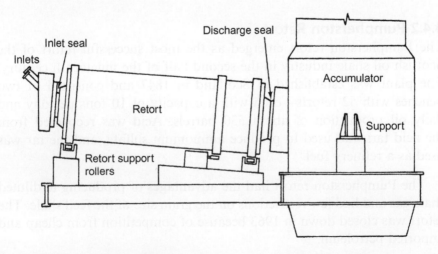

Fig. 4.6. TOSCO retort (Scouten, 1990).

gases from the ceramic ball heater—is introduced into a horizontal rotary kiln together with 1.5 times its weight in previously heated ceramic balls. The temperature of the shale is raised to its minimal retort temperature of 480°C (900°F). The kerogen is converted to distillable shale oil that is fed to a fractionator for hydrocarbon recovery and water separation. The spent shale and the ceramic balls are discharged and separated; the ceramic balls are returned to their heater; and the spent shale is cooled, moistened for dust control, and removed for land disposal. The fractionator separates the shale oil hydrocarbon vapors into gas, naphtha, gas oil, and higher boiling *bottom oil*.

The gas, naphtha, and gas oil are sent to various upgrading units, while the bottom oil is sent to a delayed coking unit, where it is converted to lower boiling products and by-product coke. Both gas oil and raw naphtha are upgraded in separate hydrogenation units through reaction with high-pressure hydrogen (produced on-site from steam reforming of the fuel gas originally recovered from the retort) (Dinneen, 1976; US OTA, 1980).

The typical spent shale produced by the TOSCO II process was a fine-grained material (containing approximately 4.5% w/w of organic carbon via char formation) comprising approximately 80 mass% of the raw oil shale feedstock. The mineral constituents of the spent shale, consisting of principally dolomite, calcite, silica, and silicates, are mostly unchanged by the retorting process treatment, except that some carbonate minerals such as dolomite have decomposed to oxides liberating carbon dioxide. During the retorting process, significant size reduction is also taking place, yielding the particle (grain) size of most spent shale finer than 8 mesh. One problem with the system was slow destruction of the ceramic balls by contact with the abrasive shale particles.

4.5 THE FUTURE

The fundamental issue with all aboveground oil shale retorting technologies is the need to provide large amounts of heat energy to thermally decompose the kerogen to liquid and gas products. More than 1 ton of shale must be heated to temperatures in the range 425–525°C (850–1000°F) for each barrel of oil generated, and the heat supplied must be of relatively high quality to reach the retorting temperature. Once the reaction is complete, recovering sensible heat from the hot rock is very desirable for optimum process economics.

This leads to three areas where new technology could improve the economics of oil recovery: (1) recovering heat from the spent shale, (2) disposal of the spent shale, especially if the shale is discharged at temperatures at which the char can catch fire in the air, and (3) concurrent generation of large volumes of carbon dioxide when the minerals contain limestone, as they do in Colorado and Utah.

Heat recovery from hot solids is not always efficient—the major exception to this generalization is in the field of fluidized bed technologies where the technologies are mature and practiced in several industries, particularly in the petroleum industry. However, applying fluidized bed technologies to oil shale would require grinding the shale to sizes less than approximately 1 mm, an energy-intensive task that would result in an increased disposal problem. Such fine particles might be used in a lower temperature process for sequestering carbon dioxide, with the costs of grinding now spread over to the solution of this problem.

Disposal of spent shale is also a problem that must be solved economically for the large-scale development of oil shale to proceed. Retorted shale contains carbon as *char*, which may represent more than 50% w/w of the original carbon value in the oil shale feedstock. The char is potentially pyrophoric and can burn if disposed into the open air while hot. In addition, the process results in a solid that occupies more volume than the fresh shale because of the problems of packing random particles. This is analogous to the tar sand (plants) in northern Alberta (Canada) where the sand from the process occupies more space than the original tar sand, an approximate 120% increase in volume (Speight, 2007, 2009).

REFERENCES

Andrews, A., 2006. Oil Shale: History, Incentives, and Policy. Report RL33359. CRS Report for Congress, Congressional Research Service, Washington, DC, April 13, 2006.

Barnet, W.I., 1982. Union Oil Company of California oil shale retorting processes. In: Allred, V.D. (Ed.), Oil Shale Processing Technology, Center for Professional Advancement, East Brunswick, NJ, pp. 169–187.

Bartis, J.T., LaTourette, T., Dixon, L., Peterson, D.J., Cecchine, G., 2005. Oil Shale Development in the United States. Report MG-414-NETL, RAND Corporation, Santa Monica, CA.

Berkovich, A.J., Young, B.R., Levy, J.H., Schmidt, S.J., 2000. Predictive heat model for Australian oil shale drying and retorting. Energy Fuels 39 (7), 2592–2600.

Burnham, A.K., McConaghy, J.R., 2006. Comparison of the acceptability of various oil shale processes. In: Proceedings of the AICHE 2006 Spring National Meeting, Orlando, FL, March 23–27, 2006.

Cashion, W.B., 1967. Geology and Fuel Resources of the Green River Formation. Professional Paper No. 548, United States Geological Survey, Washington, DC.

Dinneen, G.U., 1976. Retorting technology of oil shale. In: Yen, T.F., Chilingar, G.V. (Eds.), Oil Shale, Elsevier, Amsterdam, Netherlands, pp. 181–198.

Golubev, N., 2003. Solid heat carrier technology for oil shale retorting. Oil Shale 20 (3S), 324–332.

He, Y.G., 2004. Mining and utilization of Chinese Fushun oil shale. Oil Shale 21, 259–264.

Hohmann, J.P., Martignoni, W.P., Novicki, R.E.M., Piper, E.M., 1992. Petrosix—A successful oil shale operational complex. In: Proceedings of the Eastern Oil Shale Symposium, Kentucky, 4–11.

Hou, X.L., 1986. Shale Oil Industry in China. The Hydrocarbon Processing Press, Beijing.

Johnson, H.R., Crawford, P.M., Bunger, J.W., 2004. Strategic Significance of America's Oil Shale Resource, vol. II, Oil Shale Resources, Technology and Economics. AOC Petroleum Support Services, LLC, Washington, DC.

Kennedy, C.R., Krambeck, F.J., 1984. Surge Bin Retorting Solid Feed Material. United States Patent 4,481,100.

Kogerman, A., 1997. Archaic manner of low-temperature carbonization of oil shale in wartime Germany. Oil Shale 14, 625–629.

Koszarycz, R., Padamsey, R., Turner, L.R., Ritchey, R.M., 1991. The AOSTRA Taciuk processing—heading into the commercialization phase. In: Proceedings of the 1991 Eastern Oil Shale Symposium, Lexington, KY, November 30–December 2, 1991, 106–115.

Martignoni, W.P., Bachmann, D.L., Stoppa, E.F., Rodnignes, W.J.B., 2002. Petrosix oil shale technology learning curve. In: Symposium on Oil Shale (Abstract), Tallinn, p. 30.

Matar, S., 1982. Synfuels: Hydrocarbons of the Future. Pennwell Publishing Co., Tulsa, OK.

Matzick, A., Dannenburg, R.O., Ruark, J.R., Phillips, J.E., Lankford, J.D., Guthrie, B., 1966. Development of the Bureau of Mines Gas Combustion. Oil Shale Retorting Process. US Bureau Mines Bull. No. 635. US Bureau of Mines, Department of the Interior, Washington, DC.

Opik, J., Golubev, N., Kaidalov, A., Kann, J., Elenurm, A., 2001. Current status of oil shale processing in solid heat carrier UTT (Galoter) retorts in Estonia. Oil Shale 18, 98–108.

Schmidt, S.J., 2002. Shale oil—a path to a secure supply of oil well into this century. In: Proceedings of the Symposium on Oil Shale (Abstract and Full Paper), Tallin, Estonia, p. 28.

Schmidt, S.J., 2003. New directions for shale oil: path to a secure new oil supply well into this century (On the Example of Australia). Oil Shale 20 (3), 333–346.

Scouten, C., 1990. In: Speight, J.G. (Ed.), Fuel Science and Technology Handbook, Marcel Dekker Inc., New York.

Senchugov, K., Kaidalov, A., 1997. Utilization of rubber waste in mixture with oil shale in destructive thermal processing using the method of SHC. Oil Shale 14, 59–73.

Sonne, J., Doilov, S., 2003. Sustainable utilization of oil shale resources and comparison of contemporary technologies used for oil shale processing. Oil Shale 20 (3S), 311–323.

Speight, J.G., 2007. The Chemistry and Technology of Petroleum, fourth ed. CRC-Taylor and Francis Group, Boca Raton, FL.

Speight, J.G., 2008. Synthetic Fuels Handbook: Properties, Processes, and Performance. McGraw-Hill, New York.

Speight, J.G., 2009. Enhanced Recovery Methods for Heavy Oil and Tar Sands. Gulf Publishing Company, Houston, TX.

Speight, J.G., 2011. Handbook of Industrial Hydrocarbon Processes. Gulf Professional Publishing, Elsevier, Oxford, United Kingdom.

Speight, J.G., 2013. The Chemistry and Technology of Coal, third ed. CRC-Taylor and Francis Group, Boca Raton, FL.

Steele, H.B., 1979. The Economic Potentialities of Synthetic Liquid Fuels from Oil Shale. Arno Press Inc., New York.

Taciuk, W., 2002. The Alberta Taciuk process—capabilities for modern production of shale oil. In: Symposium on the Oil Shale (Abstract), Tallinn, Estonia, p. 27.

Taciuk, W., Turner, L.R., 1988. Development of Australian oil shale processing utilizing the Taciuk processor. Fuel 67, 1405–1407.

US DOE, 2007. Secure Fuels from Domestic Resources, the Continuing Evolution of America's Oil Shale and Tar Sands Industries: Profiles of Companies Engaged in Domestic Oil Shale and Tar Sands Resource and Technology Development. Office of Naval Petroleum and Oil Shale Reserves, Office of Petroleum Reserves, US Department of Energy, Washington, DC.

US OTA, 1980. An Assessment of Oil Shale Technologies, vol. I. Report PB80-210115. Office of Technology Assessment. Congress of the United States, Washington, DC.

Whitcombe, J.A., Vawter, R.G., 1976. The TOSCO-II oil shale process. In: Yen, T.F. (Ed.), Science and Technology of Oil Shale, Ann Arbor Science Publishers, Ann Arbor, MI. (Chapter 4).

Yefimov, Y., Doilov, S., 1999. Efficiency of processing oil shale in a 1000 ton per day retort using different arrangement of outlets for oil vapors. Oil Shale 16 (Special issue), 455–463.

Yen, T.F., 1976. Oil shales of United States—a review. In: Yen, T.F. (Ed.), Science and Technology of Oil Shale, Ann Arbor Science Publishers, Inc., Ann Arbor, MI, pp. 1–17.

Zhao, Y.H., He, Y.G., 2005. Utilization of retort gas as fuel for internal combustion engine for producing power. Oil Shale 22, 21–24.

Zhou, C., 1995. General description of Fushun oil shale retorting factory in China. Oil Shale 13, 7–11.

CHAPTER 5

In Situ Retorting

5.1 INTRODUCTION

An attractive alternative to surface retorting processes is the in situ retorting process (*underground retorting process*) (Lee et al., 2007; Scouten, 1990; Speight, 2008; US DOE, 2010, 2011). There are many reasons to develop in situ techniques for recovering oil from the oil shale. In surface processing, approximately 80% of the material mined must also be disposed of as inert inorganic matter, which presents serious environmental problems and adds considerably to the cost of the oil produced. In addition, half or more of the shale oil reserves are contained in the lower grade shale, ranging down to 10 gallons of shale oil per ton of oil shale. Only by in situ retorting can oil be economically recovered with high grade sections. The in situ retorting process involves hydraulic pressure, chemical explosives, and the more unconventional nuclear explosives. Heat is supplied either by underground combustion or by introducing heated gases or liquids to the oil shale formation.

In situ retorting is usually carried out by introducing air to burn and pyrolyze the underground oil shale layer in order to obtain shale oil (Scouten, 1990; Speight, 2008). This process obviates the problems of mining, handling, and disposing of large quantities of material, which are involved in aboveground retorting. In situ retorting offers the potential of recovering deeply buried oil shale. However, the true in situ method is not usually successful, because oil shale lacks permeability, thus hindering the inflow of air and outflow of produced oil and gases, and also lowering the heat transfer into the oil shale deposit.

In the in situ retorting process, heat is supplied either by underground combustion or by introducing heated gases or liquids to the oil shale formation. Two methods of in situ oil shale retorting have been tested. The "true" in situ method consists of fracturing, retorting, and recovering the products through the use of boreholes from the surface. The "modified" in situ method involves subsurface mining to create a void, blasting the adjacent oil shale into this void area and then retorting it.

In situ processing presents an opportunity for recovering shale oil from deep, even low-grade deposits, and the main idea in in situ combustion is to burn a portion of the oil shale in order to produce sufficient heat to retort the remainder.

In situ processes introduce heat to the kerogen while it is still embedded in the natural geological oil shale formation. There are two general approaches: (1) *true in situ processes* in which there is minimal or no disturbance of the ore bed and (2) *modified in situ processes* in which the bed is given a rubble-like texture (rubbelized), either through direct blasting with surface up-lift or after partial mining to create void space.

Both conventional and in situ retorting processes result in inefficiencies that reduce the volume and quality of the produced shale oil. Depending on the efficiency of the process, a portion of the kerogen that does not yield liquid is either deposited as *coke* (a carbonaceous residue) on the host mineral matter or is converted to hydrocarbon gases. For the purpose of producing shale oil, the optimal process is one that minimizes the regressive thermal and chemical reactions that form coke and hydrocarbon gases and maximizes the production of shale oil.

The advantages of in situ processing include the following: (1) oil can be recovered from deep deposits of oil shale formation, (2) mining costs can be eliminated or minimized, (3) issues related to solid waste are eliminated, (4) shale oil can be extracted from leaner shale, for example, deposits containing less than 15 gallons of shale oil per ton of oil shale, (5) the process is ultimately more economical because of elimination or reduction of costs involved with mining, transportation, and crushing.

However, disadvantages are also evident in in situ processing (Fang et al., 2008) and these include the following: (1) subsurface combustion is difficult to control because of insufficient permeability within the shale formation, (2) drilling costs are still high, (3) recovery efficiencies are generally low, (4) it may be difficult to establish the required permeability and porosity in the shale formation, and (5) there is the potential for aquifer contamination—if not controlled or treated, effects may linger for an extended period of time even after project completion.

Modified in situ retorting is a more amenable approach to solving the problems related to low-to-zero permeability and low-to-zero porosity.

In the process, an upper portion of the oil shale bed is taken to the surface by conventional mining to provide the desired void volume for the resulting underground retort. Then the oil shale deposit adjacent to the void portion is fractured, by using conventional explosion, to rubble, which expands to the void volume. Combustion is initiated with inflow of air at the top of the shale rubble in the underground retort, with the burning front descending through the rubble bed at the rate of several meters per week. Ahead of the combustion zone, the hot combustion gas creates a pyrolysis zone, where oil shale is thermally decomposed. The shale oil produced flows to the bottom of the rubble and is pumped to the surface.

5.2 IN SITU RETORTING

In situ retorting during the 1970s' and 1980s' oil shale boom involved dewatering to remove any groundwater, fracturing the deposit to increase permeability to fluid flow, heating the oil shale in place by injecting a hot fluid or by igniting a portion of the deposit, recovering any oil and gases produced from the heated deposit through wells, and transporting the liquid to an upgrading facility (US OTA, 1980). The most widely tested in situ retorting technology involved burning a portion of the shale underground to provide the heat necessary to retort the remaining shale. This method achieved little success owing to temperature and combustion instabilities (Bartis et al., 2005; US OTA, 1980).

However, oil shale is a relatively hard impermeable formation through which fluids will not flow is of extreme importance in the in situ processing option. The specific gravity of the Green River kerogen is approximately 1.05 and the mineral fraction has an approximate value of 2.7 (Baughman, 1978).

In situ processes can be technically feasible where permeability of the rock exists or can be created through fracturing. The target deposit is (1) fractured, (2) air is injected, (3) the deposit is ignited to heat the formation, and (4) the resulting shale oil is moved through the natural or man-made fractures to production wells that transport it to the surface, and hence to the refinery (Bartis et al., 2005). However, difficulties in controlling the flame front and the flow of the produced shale oil can limit the ultimate oil recovery, leaving portions of the deposit unheated and portions of the shale oil unrecovered.

Although in situ processes avoid the need to mine the shale, they do require that heat be supplied underground and that product be recovered from a relatively nonporous bed. As such, the in situ processes tend to operate slowly, a behavior that the Shell in situ process exploits by heating the resource to approximately 343°C (650°F), resulting in high yields of liquid products, with minimal secondary reactions (Karanikas et al., 2005; Mut, 2005).

The process involves use of ground-freezing technology to establish an underground barrier (*freeze wall*) around the perimeter of the extraction zone. The freeze wall is created by pumping refrigerated fluid through a series of wells drilled around the extraction zone. The freeze wall prevents groundwater from entering the extraction zone and keeps hydrocarbons and other products generated by the in situ retorting from leaving the project perimeter.

In situ processes avoid the spent shale disposal problems because the spent shale remains where it is created but, on the other hand, the spent shale will contain uncollected liquids that can leach into ground water, and vapors produced during retorting can potentially escape to the aquifer (Karanikas et al., 2005).

Modified in situ processes are designed to improve performance by exposing more of the target deposit to the heat source and by improving the flow of gases and liquid fluids through the rock formation and increasing the volumes and quality of the oil produced. These processes involve mining beneath the target oil shale deposit before heating and also require drilling and fracturing the target deposit above the mined area to create void space of 20–25%, which is needed to allow heated air, produced gases, and produced shale oil to flow toward production wells.

The basic premise of modified in situ processes is improved performance by exposing more of the target deposit to the heat source, improving the flow of gases and liquid fluids through the rock formation, and increasing the volumes and quality of the oil produced. To do this, the oil shale is heated by igniting the top of the target deposit. Condensed shale oil that is thermally decomposed ahead of the flame is recovered from beneath the heated zone and pumped to the surface.

The Occidental vertical modified in situ process was developed specifically for the deep, thick shale beds of the Green River Formation. Approximately 20% percent of the shale in the retort area is mined; the balance is then carefully blasted using the mined out volume to permit expansion and uniform distribution of void space throughout the retort (Hulsebos, 1988; Petzrick, 1995).

In this process, some of the shale was removed from the ground and the remainder was explosively shattered to form a packed bed reactor within the mountain. Drifts (horizontal tunnels into the mountain) provided access to the top and bottom of the retort. The top of the bed was heated with burners to initiate combustion and a slight vacuum pulled on from the bottom of the bed to draw air into the burning zone and withdraw gaseous products. Heat from combustion retorted the shale below, and the fire spread to the char left behind. Key to the success of the process is the formation of shattered shale of relatively uniform particle size in the retort.

If the oil shale contains a high proportion of dolomite (a mixture of calcium carbonate and magnesium carbonate; $CaCO_3.MgCO_3$) as occurs in Colorado oil shale, the limestone decomposes at the customary retorting temperatures to release large volumes of carbon dioxide.

$$CaCO_3.MgCO_3 \rightarrow CaO + MgO + 2CO_2$$

This consumes energy and leads to the additional problem of needing to sequester the carbon dioxide to meet global climate change concerns.

The development of a commercial oil shale industry in the United States would have significant social and economic impacts on local communities. Other impediments to such development include the relatively high cost of producing oil from oil shale and the overall lack or uniformity of regulations to lease oil shale tracts.

Hydrocarbon products of successful in situ heating are similar in character to the products recovered from aboveground retorts: petroleum gases, hydrocarbon liquids, and char. Field experiences with the first generation in situ retorts indicate that the petroleum gases tend to be of poorer quality than gases recovered by aboveground retorts. The condensable liquid fraction, however, generally tends to be of better quality than the liquid hydrocarbon fractions recovered from above-ground retorts with higher degrees of cracking of the kerogen macro-molecules and elimination of substantial portions of the higher boiling fractions typically produced in aboveground retorts.

Overall yields from any in situ retorting tend to be lower than those from equal amounts of oil shale of equivalent richness processed by aboveground retort. Various explanations have been advanced for these observed differences. Some of the loss of quality for recovered gases may be due to the dilution that results when heat is introduced to the formation by injection of combustion gases and steam, by advancement of a flame front as a result of combustion of some portion of the shale, or when high-pressure gases are used to sweep retorting products from the formation to recovery wells.

The quality improvements for the liquid fraction may be due to the relatively slow and more even heating that can be attained in a properly designed and executed in situ retorting process. Such quality improvements also may be indicative of further refining of initial retorting products when sweep gases such as natural gas or hydrogen are used. Finally, and importantly from an environmental perspective, the char and the mineral fraction to which it is adsorbed are not recovered but remain in the formation, significantly reducing (but not completely eliminating) environmentally related disposal issues.

The overall success of any in situ retorting technology results from the ability of the technology to distribute heat evenly throughout the formation. Indiscriminate formation heating that allows portions of the formation to reach 590°C (1100°F) can result in technological problems, the thermal decomposition of mineral carbonates and the formation and release of carbon dioxide.

From an operational standpoint, such decompositions are endothermic and will result in the energy demands of such uncontrolled in situ retorting quickly becoming insurmountable. As noted earlier, the environmental consequences of carbonate decomposition during in situ retorting can be expected to be mitigated to a large extent by the natural carbon dioxide sequestration that can also be anticipated. Nevertheless, a lack of precise heat control will devastate both the yields and the quality of recovered hydrocarbons and must be avoided. However, in situ retorting with good thermodynamic control can produce pyrolysis products of equal or even greater quality than aboveground retort.

Another potential disadvantage to in situ retorting involves the time that it takes to heat substantial masses of formation materials to retorting temperature (on the order of months or years) and the energy costs over that period. Field experiences are limited, and because every formation accepts heat differently, it is difficult to define a universal time line or perform precise, reliable energy balances, except on a site-specific basis.

Other largely unanswered questions involve long-term impacts from retorted segments of oil shale formations.

5.2.1 American Shale Oil, LLC

American Shale Oil LLC (AMSO) has developed a new process for in situ retorting of Green River oil shale (Burnham et al., 2009; US DOE, 2010, 2011). The American Shale Oil process involves the use of proven oil field drilling and completion practices coupled with a unique heating and recovery technology.

The process involves the use of a closed-loop in situ retorting process with advantages of energy efficiency and manageable environmental impacts. The oil shale is heated with superheated steam or other heat transfer medium through a series of pipes placed below the bed to be retorted. Shale oil and gas are produced through wells drilled vertically from the surface and *spidered* to provide a connection between the heating wells and production system. Convection and refluxing are mechanisms that improve heat transfer to retort the oil shale.

After initial start-up, the process uses the gas produced from retorting to supply all the heat required to extract the shale oil and gas from the deposit. Surface disturbance is minimized with this process by heating through lateral piping. Energy efficiency is optimized by recovery of heat from the shale rock after retorting is completed (US DOE, 2007).

5.2.2 Chevron Process

The Chevron technology for the recovery and upgrading of oil from shale (CRUSH) process is an in situ conversion process, which involves the application of a series of fracturing technologies that rubbelize the formation, thereby increasing the surface area of the exposed kerogen. The exposed kerogen in the fractured formation is then converted through a chemical process, resulting in the kerogen changing from a solid material to a liquid and gas. The hydrocarbon fluids are then recovered and improved to refinery feedstock specifications (US DOE, 2007).

In the process, heated carbon dioxide is used to heat the oil shale. Vertical wells are drilled into the oil shale formation and horizontal fractures are induced by injecting carbon dioxide through drilled wells. This is then forced under pressure through the formation for circulation through the fractured intervals to rubbelize the production zone. For further rubbelization, propellants and explosives may be used. The used carbon dioxide can then be routed to the gas generator to be reheated and recycled. The remaining organic matter in previously heated and depleted zones is combusted in situ to generate the heated gases required to process successive intervals. These gases would then be pressured from the depleted zone into the newly fractured portion of the formation and the process would be repeated. The hydrocarbon fluids are brought up in conventional vertical oil wells (US DOE, 2007, 2010).

5.2.3 Dow Process

The Dow Chemical Company's interest in oil shale was directed in the 1950s toward in situ processes and local shales, specifically the Antrim, a black Devonian shale formed some 260 million years ago during the Devonian and Mississippian ages and which underlies most of the lower peninsula of Michigan. The shale occasionally produces natural gas, which yields 9–10 gallons of retorted shale oil per ton of oil shale when retorted. The Michigan Antrim shale is believed to contain an equivalent hydrocarbon volume of 2500 billion barrels. Even applying a 10% w/w recovery factor, this resource is about nine times the amount of the U.S. proven oil reserve.

Various field studies done by Dow on the Antrim shale have resulted in the drilling of some 21 wells and the cutting of over 5000 feet of core. Both high pressure air and oxygen have been used underground as reactants. Hydraulic fracturing and chemical explosives have been used on a massive scale in attempts to generate the necessary fracture permeability for in situ retorting. The costs have been large and the technological problems yet to be solved are formidable; however, the potential of the eastern oil shale is so large that a continuing effort has been recommended to the Federal Government.

As early as June 1955, research planners for the Dow Chemical Company were suggesting that underground processes, including in situ retorting, be studied with specific attention to the essential problem of rubbelizing the strata, because the long range availability of energy and organic raw materials was not favorable at some Dow plant sites.

A block of Colorado oil shale was purchased, however, interest soon shifted to a local oil shale deposit which was located a little more than half a mile below the Dow power plants at Midland, Michigan. This shale, the Antrim of Michigan, is a Chattanooga-type shale that represents only a small part of the very extensive oil shale deposits of Devonian age in the eastern and mid-eastern parts of the United States. The U.S. Geological Survey estimates that these shales underlie an area of over 400 000 square miles.

Thus, the Dow Chemical Company, under a contract with the U.S. Department of Energy, conducted a 4-year research program to test the feasibility of deep, in situ recovery of low-heat content gas from Michigan Antrim Shale (Matthews and Humphrey, 1977; McNamara and Humphrey, 1979).

In the process, extensive fracturing (rubbelizing) of the oil shale is considered essential for adequate in situ retorting and recovery of energy from the Antrim shale. Two wells were explosively fractured using metalized ammonium nitrate slurry. The test facility was located 75 miles northeast of Detroit, Michigan, over one acre of field—the process used was true in situ (TIS) retorting.

Combustion of the shale was started using a 440-V electric heater (52 kW) and a propane burner (250 000 Btu/hr). The special features of the process include shale gasification and severe operating conditions. Their tests also showed that explosive fracturing in mechanically under-reamed wells did not produce extensive rubbelization. They also tested hydrofracturing, chemical underreaming, and explosive underreaming.

5.2.4 ExxonMobil Electrofrac Process

ExxonMobil's Electrofrac Process uses a method of heating oil shale in situ based on established principles of oil and gas well horizontal drilling and hydraulic fracturing. The proppant used in fracturing is highly electrically conductive and is a means of extending heat beyond the borehole and into the fractures. The proppant creates an electrically conductive pathway along the length of the fracture system, turning it into a large, plate-like electrode. ExxonMobil has conducted a 100-foot scale test (which it calls "The Giant Toaster") at its Colony Mine near Parachute, Colorado.

The process uses a series of hydraulic fractures created in the oil shale formation. The fractures should be preferably longitudinal vertical fractures created from horizontal wells and should conduct electricity from the heel to the toe of each heating well. For conductibility, an electrically conductive material such as calcined petroleum coke is injected into the wells through fractures, forming a heating element. Heating wells are placed in a parallel row with a second horizontal well intersecting them at their toe. This allows opposing electrical charges to be applied at either end. Planar heaters require fewer wells than wellbore heaters and offer a reduced surface footprint. The shale oil is extracted by separate dedicated production wells (Plunkett, 2008; Symington et al., 2006, 2008; US DOE, 2007).

5.2.5 Occidental Modified In Situ Process

Starting in 1972, Occidental Oil Shale, Inc., a subsidiary of Occidental Petroleum, began research on a shale oil extraction process, ending research in 1991. The company conducted the first modified in situ oil shale experiment in 1972 at Logan Wash, Colorado.

Modified in situ processes attempt to improve performance by exposing more of the target deposit to the heat source, by improving the flow of gases and liquid fluids through the rock formation, and by increasing the volumes and quality of the oil produced. Modified in situ processes involve mining beneath the target oil shale deposit before heating. It also requires drilling and fracturing of the target deposit above the mined area to create void space of 20–25%. This void space is needed to allow heated air, produced gases, and pyrolyzed shale oil to flow toward production wells. The shale is heated by igniting the top of the target deposit. Condensed shale oil that is pyrolyzed ahead of the flame is recovered from beneath the heated zone and pumped to the surface.

The Occidental vertical modified in situ process was developed specifically for the deep, thick shale beds of the Green River Formation. About 20% of the shale in the retort area is mined; the balance is then carefully blasted using the mined out volume to permit expansion and uniform distribution of void space throughout the retort. A combustion zone is started at the top of the retort and moved down through the shale rubble by management of combustion air and recycled gases. Full-scale retorts would contain 350 000 cubic meters of shale rubble (Petzrick, 1995).

In a modified in situ retorting method, a portion of the underground shale was mined and then the remaining portion was crushed through a series of explosions. This method overcame many of the difficulties of burning shale underground by allowing the necessary combustion air to permeate the crushed shale. The underground shale was then retorted in place and the mined shale was sent to surface retorts for processing. There were several companies interested in this technology in the early 1980s, but that interest faltered when oil prices collapsed (Bartis et al., 2005; US OTA, 1980).

The process used explosives to create underground chambers (retorts) of fractured oil shale. About 20% was mined out, after which blasting was used to fracture oil shale. The commercial-sized retort covered 333 by 166 feet area and had a height of 400 feet. Oil shale was then ignited on the top by external fuel, and air or steam was injected to control the process. As a result, combustion moved from the top to the bottom of the retort (US DOE 2004a, b, c, 2007).

The retorting technology involved creating a void in the oil shale formation using conventional underground mining techniques. Explosives (ammonium nitrate) and fuel oil were then introduced to cause the rubbelizing of some of the shale on the walls of the void and to expand existing fractures in the formation, improving its permeability. Access to the void was sealed and a controlled mixture of air and fuel gas (or alternatively, commercial fuel such as propane or natural gas) was introduced to initiate controlled ignition of the rubbelized shale. Combustion using this external fuel continued until the rubbelized shale itself was ignited, after which external fuel additions were discontinued and combustion air continued to be provided to the void to sustain and control combustion of the shale. The resulting heat expanded downward into the surrounding formation, heating and retorting the kerogen.

Retort products collected at the bottom of the retort void and were then recovered from conventional oil and gas wells installed adjacent to the void. Careful control of combustion air or fuel mixtures was the primary control over the rate of combustion occurring in the heavily instrumented and monitored void. Once recovery of retorted oil shale products equilibrated, a portion of the hydrocarbon gases was recycled back into the void to be used as fuel to sustain in situ combustion. Two separate retorts were constructed and operated during Phase II of the project, with the last two retorts shutting down in February 1983.

Ultimately, oil recovery was equivalent to 70% of the yield predicted through Fisher assay. Design of the experiment was directed toward potential future commercial applications so that numerous such in situ retorts were operated simultaneously to demonstrate the practicability of an approach that would likely have been desirable in commercial development ventures.

Differences in approaches among modified in situ technologies center on the manner in which formation voids are formed, the shape and orientation of such voids (horizontal vs. vertical), and the actual retorting and product recovery techniques used. Retorting techniques can include controlled combustion of rubbelized shale or formation heating by alternative means such as the introduction of electromagnetic energy. Product recovery techniques have included steam leaching, chemically assisted or solvent leaching, and displacement by high-pressure gas or water injection. Some of these formation sweeping techniques can also be seen as aiding or promoting additional refining of the initial retorting products. It is beyond the scope of this summary to discuss in detail all or even a majority of the designs that have been developed (Lee, 1991).

5.2.6 Shell In Situ Conversion Process
In the early 1980s, Shell proposed the in situ conversion (ICP) method for in situ retorting (Bartis et al., 2005). The process comprised a series of underground heaters drilled into an oil shale deposit, and the field size for this method is typically one square mile. Approximately 15–25 holes are drilled per acre at a distance of 35–42 feet (10.7–12.8 m) apart in a variety of configurations. The wells reach a depth of up to 2000 feet, depending on the deposit location (Bartis et al., 2005). The target depth zone is 1000–2000 feet (Andrews, 2006). The fracturing step is achieved by existing and induced fractures—as in petroleum recovery processes, fracturing increases shale permeability (Scouten, 1990; Speight, 2007).

The electrical resistance heaters inserted into the holes reach temperatures of 760°C (1400°F) to raise the surrounding shale deposit to an average temperature of 345°C–370°C (650°F–700°F). Although this temperature is significantly lower than that required for conventional surface retorting (482°C–538°C or 900°F–1000°F), it is sufficient to induce the chemical and physical changes that release the oil from the shale.

After heating the deposit for 2–3 years, the oil and any associated gas are pumped out of the ground using conventional methods (Bartis et al., 2005). The hydrocarbon mixture generated from this procedure is of very high quality and quite different from traditional crude oils, in that it contains light hydrocarbons and almost no heavy ends. The mixture quality can be controlled by adjusting the heating time, temperature, and pressure in the subsurface shale layer. A typical mixture is two-thirds liquid (30% naphtha, 30% jet fuel, 30% diesel, and 10% heavier oil) and one-third gas (propane and butane). The liquid hydrocarbon fractions can easily be converted into a variety of finished products, including gasoline, naphtha, jet fuel, and diesel (Andrews, 2006). On a 30×40 feet (9.1×12.2 m) test area, Shell recovered 1700 barrels of light oil plus associated gas from shallower, less-concentrated oil shale layers.

To protect groundwater, a freeze wall is constructed around the heating grid at a distance of 300 feet (91 m) from the heaters. The coolant is a 40% ammonia-water mixture. The freeze wall establishes an underground barrier to fluid movement, thus preventing groundwater contamination and the escape of the freed shale oil.

Shell's calculation of 3.5 units of energy gained from the oil shale for every unit of energy consumed through the electrical heating process assumes electricity is produced by an advanced, 60% efficient, combined cycle gas power plant. Shell is also considering gas-fired heating, which will utilize the natural gas being recovered from the drilling process, potentially improving the energy balance.

5.3 THE FUTURE

In the 1970s, during the period of world oil crisis, different oil shale retorting technologies, including underground retorting, were actively developed by various oil companies. In the second half of the twentieth century, extensive research and development efforts had been devoted to the commercialization of the in situ pyrolysis of oil shale. However, most of these efforts were either halted or scaled down in the late twentieth century because of the unfavorable process economics in a short-term. In fact, until the 1990s, because of the depletion of oil price, many in situ projects were closed or plans abandoned and, even at the time of writing, although the crude oil price is much higher than decades ago, only the Shell Company in the United States is conducting an in situ oil shale retorting pilot test.

In situ retorting obviates any issues related to mining, handling, and disposing of large quantities of material, as occurs in aboveground retorting. In situ retorting offers the potential for recovering deeply deposited oil shale. However, to mitigate the potential disadvantages of in situ processes, novel and advanced retorting and upgrading processes should seek to modify the conversion chemistry to improve recovery and create high-value by-products. Approaches such as (1) lower heating temperatures, (2) higher heating rates, (3) shorter residence time, (4) use of additives such as hydrogen (or hydrogen transfer/donor agents), and (5) use of solvents need to be considered.

Several emerging technologies for extracting oil from shale are currently being tested and refined in Utah and Colorado in an effort to determine their commercial viability. These efforts range from application of new surface mining and processing technologies to modified in situ methods.

The basic idea is to create an in situ environment in an ex situ process. The shale is mined from the richest and largest deposits and then crushed and piled into an embankment lined with clays or other impervious material. The embankment is capped with an appropriate impermeable layer. Horizontal wells are drilled into the structure and then heaters are inserted into the wells to provide a slow, steady heating source. The concept is to mimic in situ conditions in this embankment structure. As the source rock heats up, appropriately placed wells are used to collect the oil. Once oil production ceases, a heat scavenging program is used to scavenge or divert waste heat to adjacent embankments. Then, the structure is shut down.

Reclamation is engineered to ensure no further environmental impact from the embankment. The advantages of the technology are that features of in situ and ex situ methods are combined to process the oil shale. In fact, it may be possible to engineer improved "in situ" conditions in the embankment to produce better quality oil. However, mining would require its own infrastructure and remediation activities and the long-term environmental security of the embankment would have to be ensured.

The main difference between using horizontal drilling and hydraulic fracturing for crude oil or natural gas and using them for oil shale is the time frame—to convert the kerogen to shale oil, heat delivered electrically via the calcined petroleum coke proppant subjects the surrounding rock to the effects of pyrolysis over a period of around 5 years in order to yield its liquid oil.

A process has been proposed to convert oil shale by radio frequency heating over a period of months to years to create a product similar to natural petroleum (Burnham, 2003). Electrodes would be placed in drill holes, either vertical or horizontal, and a radio frequency chosen so that the penetration depth of the radio waves is of the order of tens to hundreds of meters. A combination of excess volume production and overburden compaction drives the oil and gas from the shale into the drill holes, where it is pumped to the surface.

It could well be that the comparative economics of shale oil will change substantially to be more favorable in the twenty-first century.

REFERENCES

Andrews, A., 2006. Oil Shale: History, Incentives, and Policy. Report RL33359. CRS Report for Congress, Congressional Research Service, Washington, DC, April 13, 2006.

Bartis, J.T., LaTourette, T., Dixon, L., Peterson, D.J., Cecchine, G., 2005. Oil Shale Development in the United States. Report MG-414-NETL. RAND Corporation, Santa Monica, CA.

Baughman, G.L., 1978. Synthetic Fuels Data Handbook, second ed. Cameron Engineers Inc., Denver, CO.

Burnham, A.K., 2003. Slow Radio-Frequency Processing of Large Oil Shale Volumes to Produce Petroleum-Like Shale Oil. Report No. UCRL-ID-155045. Lawrence Livermore National Laboratory, US Department of Energy, Livermore, CA.

Burnham, A.K., Day, R.L., Hardy, M.P., Wallman, P.H., 2009. AMSO's novel approach to in situ oil shale recovery. In: Proceedings of the 8th World Congress of Chemical Engineering, Montreal, Quebec, Canada.

Fang, C., Zheng, D., Liu, D., 2008. Main problems in the development and utilization of oil shale and the status of in situ conversion in China. In: Proceedings of the 28th Oil Shale Symposium, Colorado School of Mines, Golden, CO, October 13–15.

Hulsebos, J., Pohani, B.P., Moore, R.E., Zahradnik, R.L., 1988. Modified-in-situ technology combined with aboveground retorting and circulating fluid bed combustors could offer a viable method to unlock oil shale reserves in the near future. In: Zhu, Y.J. (Ed.), Proceedings of the International Conference on Oil Shale and Shale Oil, China Chemical Industry Press, Beijing, pp. 440–447.

Karanikas, J.M., de Rouffignac, E.P., Vinegar, H.J., Wellington, S., 2005. In Situ Thermal Processing of an Oil Shale Formation While Inhibiting Coking. United States Patent 6,877,555, April 12, Houston, TX.

Lee, S., 1991. Oil Shale Technology. CRC Press, Taylor & Francis Group, Boca Raton, FL.

Lee, S., Speight, J.G., Loyalka, S.K., 2007. Handbook of Alternative Fuel Technologies. CRC-Taylor and Francis Group, Boca Raton, FL.

Matthews, R., Humphrey, J.P., 1977. A Search for Energy from the Antrim. Paper 6494-MS. In: Proceedings of the SPE Midwest Gas Storage and Production Symposium, Indianapolis, IN, April 13–15.

McNamara, P.H., Humphrey, J.P., 1979. Hydrocarbons from Eastern oil shale. Chemical Engineering Progress, September, 88 pp.

Mut, S., 2005. Testimony before the United States Senate Energy and Natural Resources Committee, Tuesday, April 12. http://energy.senate.gov/hearings/testimony.cfm?id=1445&wit_id=4139

Petzrick, P.A., 1995. Oil Shale and Tar Sand: Encyclopedia of Applied Physics, vol. 12. VCH Publishers Inc., New York, pp. 77–99.

Plunkett, J.W., 2008. Plunkett's Energy Industry Almanac 2009. The Only Comprehensive Guide to the Energy & Utilities Industry. Plunkett Research Ltd., Houston, TX.

Scouten, C., 1990. Part IV: Oil Shale. Chapters 25–31. In: Speight, J.G. (Ed.), Fuel Science and Technology Handbook, Marcel Dekker Inc., New York.

Speight, J.G., 2007. The Chemistry and Technology of Petroleum, fourth ed. CRC-Taylor and Francis Group, Boca Raton, FL.

Speight, J.G., 2008. Synthetic Fuels Handbook: Properties, Processes, and performance. McGraw-Hill, New York.

Symington, W.A., Olgaard, D.L., Otten, G.A., Phillips, T.C., Thomas, M.M., Yeakel, J.D., 2006. ExxonMobil's electrofrac process for in-situ oil shale conversion. Paper S05B. In: Proceedings of the 26th Oil Shale Symposium, Golden, CO, October 16–20.

Symington, W.A., Olgaard, D.L., Otten, G.A., Phillips, T.C., Thomas, M.M., Yeakel, J.D., 2008. ExxonMobil's electrofrac for in-situ oil shale conversion. AAAPG Annual Convention, San Antonio, TX.

US DOE, 2004a. Strategic Significance of America's Oil Shale Reserves, I. Assessment of Strategic Issues, March. http://www.fe.doe.gov/programs/reserves/publications

US DOE, 2004b. Strategic Significance of America's Oil Shale Reserves, II. Oil Shale Resources, Technology, and Economics, March. http://www.fe.doe.gov/programs/reserves/publications

US DOE, 2004c. America's Oil Shale: A Roadmap for Federal Decision Making, USDOE Office of US Naval Petroleum and Oil Shale Reserves. http://www.fe.doe.gov/programs/reserves/publications

US DOE, 2007. Secure Fuels from Domestic Resources: The Continuing Evolution of America's Oil Shale and Tar Sands Industries. Profiles of Companies Engaged in Domestic Oil Shale and Tar Sands Resource and Technology Development. Office of Naval Petroleum and Oil Shale Reserves, Office of Petroleum Reserves, June, US Department of Energy, Washington, DC.

US DOE, 2010. Secure Fuels from Domestic Resources: The Continuing Evolution of America's Oil Shale and Tar Sands Industries. Profiles of Companies Engaged in Domestic Oil Shale and Tar Sands Resource and Technology Development, fourth ed. Office of Naval Petroleum and Oil Shale Reserves, Office of Petroleum Reserves, June, US Department of Energy, Washington, DC.

US DOE, 2011. Secure Fuels from Domestic Resources: The Continuing Evolution of America's Oil Shale and Tar Sands Industries. Profiles of Companies Engaged in Domestic Oil Shale and Tar Sands Resource and Technology Development, fifth ed. Office of Naval Petroleum and Oil Shale Reserves, Office of Petroleum Reserves, June, US Department of Energy, Washington, DC.

US OTA, 1980. An Assessment of Oil Shale Technologies, vol. I. Report PB80-210115. Office of Technology Assessment. Congress of the United States: Washington, DC.

CHAPTER 6

Refining Shale Oil

6.1 INTRODUCTION

As the demand for precursor fractions to fuels (such as gasoline and diesel) increases, there is much interest in developing economical methods for recovering liquid distillate fraction hydrocarbons from oil shale. However, these distillates are not yet economically competitive against petroleum crude. Furthermore, the value of hydrocarbons recovered from oil shale is lessened by the presence of undesirable contaminants. The major contaminants of concern are oxygen-containing, sulfur-containing, nitrogen-containing, and metallic (and organometallic) compounds that cause detrimental effects to the catalysts used in the subsequent refining processes. These contaminants, when present in the actual fuel, are also undesirable because of their disagreeable odor, corrosive characteristics, and emissions when combusted that further cause environmental problems (Kundu et al., 2006; Lee, 1991; Lee et al., 2007; Scouten, 1990; Speight, 2007, 2008; Tsai and Albright, 2006). Oil shale distillates also have a higher content of olefins than crude oil as well as a higher pour point and viscosity.

Depending on the process, some oil shale distillates have a much higher concentration of higher boiling point compounds that would favor the production of middle distillates (such as diesel and jet fuels) rather than naphtha (Hunt, 1983; Scouten, 1990). Aboveground retorting processes tended to yield lower API gravity oil than the in situ processes, because the residence time in the underground retort encourages further cracking of the primary products. Thus, additional processing equivalent to hydrocracking would be required to convert oil shale distillates to a lighter range hydrocarbon (gasoline). Removal of sulfur and nitrogen would, however, require hydrotreating.

However, shale retorting processes produce oil with none-to-very-little high-boiling residuum (Scouten, 1990). With upgrading, shale oil can

Shale Oil Production Processes

become a low-boiling premium product that has the potential to be at least as valuable as conventional petroleum (Lee, 1991; Lee et al., 2007; Scouten, 1990; Speight, 2007, 2008). However, the properties of shale oil vary as a function of the production (retorting) process. Fine mineral matter carried over from the retorting process and the high viscosity and instability of shale oil produced by present retorting processes have necessitated upgrading it before transport to a refinery.

After fines removal, the shale oil is hydrotreated to reduce the nitrogen, sulfur, and arsenic content and improve stability; the cetane index of the diesel and heater oil portion is also improved. The hydrotreating step is generally accomplished by a fixed catalyst bed processes at high hydrogen pressures, and hydrotreating conditions are slightly more severe than for comparable boiling range petroleum stocks because of the higher nitrogen content of shale oil.

However, catalytic hydrodesulfurization processes are not always the best solution for the removal of sulfur constituents from gasoline when high proportions of unsaturated compounds are present, because a significant amount of the hydrogen would be used for hydrogenation of the unsaturated components. On the other hand, when hydrogenation of the unsaturated hydrocarbons is desirable, catalytic hydrogenation processes would be effective.

6.2 SHALE OIL PROPERTIES

Shale oil is a synthetic crude oil produced by retorting oil shale and is the pyrolysis product of the organic matter (kerogen) contained in it. The raw shale oil produced from retorting oil shale can vary in properties and composition (Lee, 1991; Scouten, 1990; Speight, 2008; US DOE, 2004a, 2004b, 2004c). The two most significant characteristics of U.S. western oil shales are their high hydrogen content, due primarily to the high concentrations of paraffins (waxes), and the high concentration of nitrogen, derived from high concentrations of pyridines and pyrroles.

Compared with petroleum, shale oil is high in nitrogen and oxygen compounds and also has a higher specific gravity—in the order of 0.9–1.0 owing to the presence of heavy nitrogen-, sulfur-, and oxygen-containing compounds. Shale oil also has a relatively high pour point, and small quantities of arsenic and iron are present.

The presence of the polar constituents (containing nitrogen and oxygen functions, sulfur compounds are less of a problem) in the shale oil can be notoriously incompatible with conventional petroleum feedstocks and with petroleum products (Mushrush and Speight, 1995; Speight, 2007). As a result, particular care must be taken to ensure that all of the compounds that cause such incompatibility are removed from the shale oil before it is blended with a conventional petroleum.

6.3 HYDROCARBONS

The fundamental structure of the organic matter in oil shale gives rise to significant quantities of waxes consisting of long-chain normal alkanes which are distributed throughout the raw shale oil. However, the composition of shale oil depends on the shale from which it was obtained, as well as on the retorting method by which it was produced (Table 6.1) (Lee, 1991; Scouten, 1990; Speight, 2008). As compared with petroleum crude, shale oil is heavy, viscous, and is high in nitrogen and oxygen compounds.

Of the possible variables that can affect the quality of shale oil, the retorting method is by far the most significant. The major difference in shale oils that are produced by different processing methods is in the boiling point distribution. Rate of heating as well as temperature level and duration of product exposure to high temperature affect product type and yield.

Retorting processes which use flash pyrolysis produce more fragments containing high-molecular-weight, multi-ring aromatic structures. Processes that use slower heating conditions, with longer reaction times at

Table 6.1. Properties of Shale Oil from Various Sources

Location	Sp gr (API)	Elemental Analysis (% w/w wt. %)					Analysis of Distillate (% w/w) < 350°C		
		C	H	0	N	S	Saturates	Olefins	Aromatics
Colorado, USA	0.943 (18.6)	84.90	11.50	0.80	2.19	0.61	27	44	29
Kukersite, Estonia	1.010	82.85	9.20	6.79	0.30	0.86	22	25	53
Stuart, Australia		82.70	12.40	3.34	0.91	0.65			
Rundle, Australia	0.636 (0.91)	79.50	11.50	7.60	0.99	0.41	48	2	50
Irati, Brazil	0.919 (22.5)	84.30	12.00	1.96	1.06	0.68	23	41	36
Maoming, China	0.903	84.82	11.40	2.20	1.10	0.48	55	20	25
Fushun, China	0.912	85.39	12.09	0.71	1.27	0.54	38	37	25

low temperatures 300–400°C (570–750°F), tend to produce higher concentrations of n-alkanes. Naphthene aromatic compounds of intermediate boiling range (such as 200–400°C, 390–750°F) also tend to be formed by the slower heating processes.

Saturated hydrocarbons in the shale oil include n-alkanes, iso-alkanes, and cycloalkanes, and the alkenes consist of n-alkcnes, iso-alkenes, and cycloalkenes, while the main components of aromatics are monocyclic, bicyclic, and tricyclic aromatics and their alkyl-substituted homologues. There is a variation of the distribution of saturated hydrocarbons, alkenes, and aromatics in the different boiling ranges of the shale oil product. Saturated hydrocarbons in the shale oil increase and the aromatics increase slightly with a rise in boiling range, while alkenes decrease with a rise in boiling range.

A typical Green River shale oil contains 40% w/w hydrocarbons and 60% w/w heteroatomic organic compounds, which contain nitrogen, sulfur, and oxygen. The nitrogen occurs in ring compounds with nitrogen in the ring; for example, pyridines, pyrroles, and nitriles, and typically comprises 60% w/w of the heteroatomic organic components. Another 10% w/w of these components are organic sulfur compounds, including thiophenes as well as sulfides and disulfides. The remaining 30% w/w are oxygen-containing compounds, which occur as phenols and carboxylic acids.

Shale oil contains a large variety of hydrocarbon compounds (Table 6.2) but it also has a higher nitrogen content than the 0.2–0.3% w/w of a

Table 6.2. Major Compound Types in Shale Oil	
Saturates	Heteroatom Systems
Paraffins	Benzothiophenes
Cycloparaffins	Dibenzothiophenes
Olefins	Phenols
Aromatics	Carbazoles
Benzenes	Pyridines
Indans	Quinolines
Tetralins	Nitriles
Napthalenes	Ketones
Biphenyls	Pyrroles
Phenanthrenes	
Chrysenes	

Table 6.3. Challenges for Oil Shale Processing	
Parameters	Plugging on Processing
	Product quality
Arsenic content	Toxicity
	Catalyst poison
High pour point	Oil not pipeline quality
Nitrogen content	Catalyst poison
	Contributes to instability
	Toxicity
Diolefins	Contributes to instability
	Plugging on processing

typical petroleum (Scouten, 1990; Speight, 2007, 2008). In addition, shale oil also has a high olefin content and diolefin content—constituents which are not present in petroleum and which require attention during processing due to their tendency to polymerize and form gums and sediments (fuel line deposits). It is the presence of these olefins and diolefins, in conjunction with the high nitrogen content, which makes shale oil characteristically difficult to refine (Table 6.3). Crude shale oil also contains appreciable amounts of arsenic, iron, and nickel that also interfere with refining.

Other characteristic properties of shale oils are (1) high levels of aromatic compounds, deleterious to kerosene and diesel fractions; (2) low hydrogen-to-carbon ratio; (3) low sulfur levels, compared with most crudes available in the world (although for some shale oils from the retorting of marine oil shale, high sulfur compounds are present); (4) suspended solids (finely divided rock) typically due to entrainment of the rock in the oil vapor during retorting; and (5) low-to-moderate levels of metals.

Because of these characteristics, further processes are needed to improve the properties of shale oil products. The basic unit operations in the oil refining are distillation, coking, hydrotreating, hydrocracking, catalytic cracking, and reforming. The process selected will largely depend on the availability of equipment and the individual economics of the particular refinery.

Although the content of asphaltene constituents or resin constituents in shale oil is low—shale oil being a distillate—asphaltene constituents in shale oil may be unique, since shale oil has a high heteroatomic content that causes precipitation as an asphaltene fraction, rather than this occurring because of high molecular weight—for example, the hydroxypyridine derivatives are insoluble in low-molecular-weight alkane solvents. The polarity of the nitrogen polycyclic aromatic constituents may also explain the specific properties of emulsification of water and metal complexes.

6.4 NITROGEN-CONTAINING COMPOUNDS

Nitrogen compounds in shale oil present technological difficulties in the downstream processing of shale oil, particularly poisoning of the refining catalysts. Such compounds all originate from the oil shale and their amount and types depend heavily on the geochemistry of oil shale deposits. Since direct analysis and determination of molecular forms of nitrogen-containing compounds in oil shale rock is very difficult, the analysis of shale oil that is extracted by retorting processes provides valuable information regarding the organo-nitrogen species originally present in the oil shale.

The nitrogen content of the shale oil is higher than natural crude oil (Guo and Qin, 1993; Hunt, 1983; Scouten, 1990; Speight, 2007, 2008). The nitrogen-containing compounds identified in shale oils can be classified as basic, weakly basic, and nonbasic.

The basic nitrogen compounds in shale oils are pyridine, quinoline, acridine, amine, and their alkyl-substituted derivatives, the weakly basic ones are pyrrole, indole, carbazole, and their derivatives, and the nitrile and amide homologues are the nonbasic constituents.

Most of these compounds are useful chemicals (Scouten, 1990), although some of them are believed to affect the stability of shale oil. Generally, basic nitrogen accounts for about one-half of the total nitrogen and is evenly distributed in the different boiling point fractions. Nitrogen compounds occur throughout the boiling ranges of the shale oil, but tend to occur more in the high boiling point fractions. Levels of pyrrole-type nitrogen increase with a rise in the boiling point of the shale oil fractions. Porphyrins may occur in the high boiling point fraction of the shale oil.

Of the nitrogen-containing compounds in the <350°C (<660°F) light shale oil fraction, the majority contain one nitrogen atom. Benzoquinoline derivatives, principally acridine and alkyl-substituted homologues, could not be present significantly in the lower-boiling shale oil fractions because the boiling point of benzoquinoline and its alkyl-substituted homologues is higher than 350°C (660°F).

Organic nitrogen-containing compounds in the shale oil poison the catalysts used in different catalytic processes. They also contribute to stability problems during storage of shale oil products since they induce polymerization processes, which cause an increase in the viscosity and give rise to the odor and color of the shale oil product. The high nitrogen content of shale oil could contribute to the surface and colloidal nature of shale oil, which forms emulsions with water.

6.5 OXYGEN-CONTAINING COMPOUNDS

The oxygen content of shale oil is much higher than that of natural petroleum. Low-molecular-weight oxygen compounds in shale oil are mainly phenolic constituents—but carboxylic acids and nonacidic oxygen compounds such as ketones are also present. Low molecular weight phenolic compounds are the main acidic oxygen-containing compounds in the low-boiling fraction of the shale oil and are usually derivatives of phenol, such as cresol and polymethylated phenol derivatives.

The oxygen content of petroleum is typically in the order of 0.1– 1.0% w/w whereas those of shale oils are much higher and vary with different shale oils (Scouten, 1990; Speight, 2007). In addition, the oxygen content varies with boiling point fraction. In general, it increases as the boiling point increases, and most of the oxygen atoms are concentrated in the high boiling point fraction.

Other oxygen-containing constituents of shale oil include small amounts of carboxylic acids and nonacidic oxygen-containing compounds with a carbonyl functional group such as ketones, aldehydes, esters, and amides, which are generally present in the <350°C (<660°F) fraction. Ketones in the shale oil mainly exist as 2- and 3-alkanones. Other oxygen-containing compounds in the low-boiling (<350°C, <660°F) fraction include alcohols, naphthol, and ether constituents.

6.6 SULFUR COMPOUNDS

Sulfur compounds in the shale oils include thiols, sulfides, thiophenes, and other miscellaneous compounds. Elemental sulfur is found in some crude shale oils but is absent in others.

Generally, the sulfur content of oil shale distillates is comparable in weight percentage to that of crude oil (Lee, 1991; Lee et al., 2007; Scouten, 1990; Speight, 2007, 2008). Refiners will able to meet the current 500 ppm requirement by increasing the existing capacity of their hydrotreatment units and adding new units. However, refineries may face difficulty in treating diesel to less than 500 ppm. The remaining sulfur is present in multi-ring thiophene-type compounds that prove difficult to hydrotreat because the molecular ring structure attaches the sulfur on two sides and, if alkyl groups are present, provides steric protection for the sulfur atom. Although these compounds occur throughout the range of petroleum distillates, they are more concentrated in the residuum.

6.7 SHALE OIL UPGRADING

By comparison, a typical 35° API gravity crude oil may be composed of up to 50% of gasoline and middle-distillate range hydrocarbons. West Texas Intermediate crude (benchmark crude for trade in the commodity futures market) has 0.3% sulfur by weight, and Alaska North Slope crude has 1.1% sulfur by weight. The New York Mercantile Exchange (NYMEX) specifications for light "sweet" crude limits the sulfur content to 0.42% or less (ASTM D4294), and requires an API gravity between 37° and 42° (ASTM D287).

The first step is to prepare the crude shale oil for refining. As the oil exits the retort, it is by no means a pure distillate.

Crude shale oil usually contains emulsified water and suspended solids. Therefore, the first step in upgrading is usually dewatering and desalting. Furthermore, if not removed, the arsenic and iron in shale oil would poison and foul the supported catalysts used in hydrotreating. Because these materials are soluble, they cannot be removed by filtration. Several methods have been used specifically to remove arsenic and iron. Other methods involve hydrotreating; these also reduce the sulfur, olefin, and diolefin contents and thereby make the upgraded product less prone to gum formation. After these steps, the shale oil may be suitable for admittance to typical refinery processing.

A conventional refinery distills crude oil into various fractions according to boiling point range, before further processing (Speight, 2007). In order of their increasing boiling range and density, the distilled fractions are fuel gases, light and heavy straight-run naphtha (32–200°C; 90–390°F), kerosene (200–270°C; 390–520°F), gas oil (270–565°C; 520–1050°F), and residuum (565°C+; 1050°F+) (Speight, 2007). Crude oil may contain 10–40% naphtha, and early refineries directly distilled a low-boiling naphtha (which is still incorrectly referred to as straight-run *gasoline*) of low-octane rating. A typical *cracking refinery* may convert a barrel of crude oil into high yields of naphtha (from which gasoline is produced) and lower yields of other distillate fuels such as kerosene (from which jet fuel and diesel fuel are produced), depending on the refinery configuration, the slate of crude oils refined, and the seasonal product demands of the market.

Hydrotreating is the option of choice to produce a stable product that is comparable to benchmark crude oils (Andrews, 2006; Speight, 2007, 2008). In terms of refining and catalyst activity, the nitrogen content of shale oil is a disadvantage. But in terms of the use of shale oil residua as a modifier for asphalt, where nitrogen species can enhance binding with the inorganic aggregate, the nitrogen content can be beneficial. If not removed, the arsenic and iron in shale oil would poison and foul the supported catalysts used in hydrotreating (Speight, 2007).

Catalytic processes that consume hydrogen are used to saturate olefins, to eliminate heterocyclic compounds (containing atoms of O, N, S), and stabilize oils to reduce tendency for oxidation and gum formation as a result of exposure to air and temperatures. Schemes known in petrochemistry to produce motor fuels of high quality from crude oil cannot be used with shale oils due to the wide boiling range of heteroatomic compounds, which means that they are present not only in the heavy fractions but also in the lighter ones.

The moving-bed hydroprocessing reactor is used to produce crude oil from oil shale or tar sands containing large amounts of highly abrasive particulate matter such as rock dust and ash. Hydroprocessing takes place in a dual-function moving-bed reactor, which simultaneously removes particulate matter by the filter action of the catalyst bed. The effluent from the moving-bed reactor is then separated and further hydroprocessed in fixed-bed reactors with fresh hydrogen added to the heavier hydrocarbon fraction to promote desulfurization.

Clay-based metallic catalysts help transform complex hydrocarbons to lighter molecular chains in modern refining processes. The *catalytic cracking* process developed during the World War II era enabled refineries to produce high-octane gasoline needed for the war effort. *Hydrocracking*, which entered commercial operation in 1958, improved on catalytic cracking by adding hydrogen to convert residuum into high-quality motor gasoline and naphtha-based jet fuel. Many refineries rely heavily on *hydroprocessing* to convert low-value gas oils residuum to high-value transportation fuel demanded by the market.

Middle-distillate range fuels (diesel and jet) can be blended from a variety of refinery processing streams. To blend jet fuel, refineries use desulfurized straight-run kerosene, kerosene boiling-range hydrocarbons from a hydrocracking unit, and light coker gas oil (cracked residuum). Diesel fuel can be blended from naphtha, kerosene, and low-boiling cracked oil from coker and fluid catalytic cracking units. From the standard 42-gallon barrel of crude oil, U.S. refineries may actually produce more than 44 gallons of refined products through the catalytic reaction with hydrogen.

Upgrading crude shale oil is necessary to reduce sulfur and nitrogen levels and contamination by mineral particulates. Since large portions of nitrogen and sulfur species in shale oil are present as heteroaromatic constituents, there is the need for selective removal of heteroaromatics, final product quality control, and molecular weight reduction.

Raw shale oil has the relatively high pour point of 24–27°C (75–80°F), compared with −24°C (−30°F) for Arabian light crude oil. Olefins and diolefins may account for as much as one-half of the low-boiling fraction of 315°C (600°F) or lower and lead to the formation of gums.

Upgrading activities are dictated by factors such as the initial composition of the oil shale, the compositions of retorting products, the composition and quality of desired petroleum feedstocks or petroleum end products of market quality, and the business decision to develop other by-products such as sulfur and ammonia into saleable products. Product variety and quality issues aside, there are other logistical factors that determine the extent to which upgrading activities are conducted at the mine site. Most prominent among these factors is the ready availability of electric power and process water. In especially remote locations, factors such as these represent the most significant parameters in mine site upgrading decisions.

The initial composition of the crude shale oil produced in the retorting step is the primary influence in the design of the subsequent upgrading operation. In particular, nitrogen compounds, sulfur compounds, and organometallic compounds dictate the upgrading process that is selected. In general, crude shale oil typically contains nitrogen compounds (throughout the total boiling range of shale oil) in concentrations that are 10–20 times the amounts found in typical crude oils. Removal of the nitrogen-bearing compounds is an essential requirement of the upgrading effort since nitrogen is poisonous to most catalysts used in subsequent refining steps and creates unacceptable amounts of NO_x pollutants when nitrogen-containing fuels are burned.

Sulfur, also a poison to refinery catalysts, is typically present in much lower proportions as organic sulfides and sulfates. With respect to sulfur, crude shale oil compares favorably with most low-sulfur crude oils, which are preferred feedstocks for low-sulfur fuels that are often required by local air pollution regulations. Hydrotreating to the extent necessary to convert nitrogen compounds to ammonia is sufficient in most instances to simultaneously convert sulfur to hydrogen sulfide. Crude shale oil additionally contains much higher amounts of organometallic compounds than conventional crude oils. The presence of these organometallic compounds complicates the mine site upgrading since they can readily foul the catalysts used in hydrotreating, causing interruptions in production and increased volumes of solid wastes requiring disposal, sometimes even requiring specialized disposal as hazardous wastes because of the presence of spoiled heavy metal catalysts.

Desired end products for mine site upgrading are typically limited to mixtures of organic compounds that are acceptable for use as conventional refinery feedstock; however, it is possible to produce feedstocks that are of higher quality and value to refineries than even crude oils with the most desirable properties. Since crude shale oils are typically more viscous than conventional crude oils, their yields of lighter distillate fractions such as gasoline, kerosene, jet fuel, and diesel fuel are typically low. However, additional hydrotreating can markedly increase the typical yields of these distillate fractions.

Raw shale oil typically contains 1.5–2.0% w/w nitrogen, 0.5–1.0% w/w oxygen, and 0.1–1.0% w/w sulfur. Sulfur and nitrogen removal is necessary and must be complete since these compounds poison most of the catalysts used in refining, and sulfur oxides (SO_x) and nitrogen oxides (NO_x) are highly publicized air pollutants.

Since raw shale oil is a condensed overhead product of pyrolysis, it does not contain the same kinds of macromolecules found in petroleum residua and coal tar or pitch (Speight, 2007, 2013). Conventional catalytic cracking, however, is an efficient technique for molecular weight reduction. It is crucially important to develop a new cracking catalyst that is more resistant to basic poisons (nitrogen and sulfur compounds) (Kundu et al., 2006). At the same time, research should also focus on reduction of the molecular weight of shale oil crude with low consumption of hydrogen.

Upgrading technologies are classified as primary, secondary, or enhanced. Primary upgrading is mainly a molecular weight reduction process (especially in the case of tar sand bitumen; Speight, 2007, 2008), while secondary upgrading involves removal of impurities from the feed (which is more pertinent to upgrading synthetic crude oil from oil shale). The primary upgrading processes may or may not use a catalyst, while the secondary processes are always catalytic.

Lower quality synthetic crude oil, such as shale oil from surface retorts, produces lower quantities of conventional refinery products than light crude oils. As a result, the value of these oils is less than that of many higher API crude oils. Upgrading is the process of converting these lower value oils to products more suitable for conventional refinery feedstocks. Partial upgrading reduces the amount of heteroatoms and other refinery-objectionable constituents to render the shale oil suitable for transportation to a refinery (Lee et al., 2007; Scouten, 1990). In contrast, the ICP process produces a refinery-ready shale oil that will not require partial upgrading prior to transportation to a refinery. The most common international standard for upgrading is the conversion of the vacuum residue to lower boiling point fractions.

Before World War II, an industrial process for shale oil upgrading was worked out—retorting of oil shale was conjunct with cracking of shale oil in vapor phase. Furthermore, it was established that cracking of shale oil in the presence of lime (as well as shale ash containing lime and aluminosilicates) resulted in an increase of the yield of gasoline whereas a lot of phenolic oxygen is bound into carbonic acid and so great deal of hydrogen is released to form gasoline. Also, resin donors are stabilized and H_2S is formed at 400–500°C (750–930°F).

It is known that heavier hydrocarbon fractions and refractory materials can be converted into lighter materials by hydrocracking. Hydrocracking processes are most commonly used on liquefied coals or heavy residual or distillate oils for the production of substantial yields of low boiling saturated products and to some extent of intermediates that are used as domestic fuels and still heavier cuts, which find uses as lubricants. These destructive hydrogenation or hydrocracking processes may be operated on a strictly thermal basis or in the presence of a catalyst. Thermodynamically speaking, larger hydrocarbon molecules are broken into lighter species when subjected to heat. The hydrogen-to-carbon (H/C) ratio of such molecules is lower than that of saturated hydrocarbons and an abundant supply of hydrogen improves this ratio by saturating reactions, thus producing liquid species. These two steps may occur simultaneously.

However, the application of the hydrocracking process has been hampered by the presence of certain contaminants in such hydrocarbons. The presence of sulfur- and nitrogen-containing compounds along with organometallics in crude shale oils and various refined petroleum products has long been considered undesirable. Desulfurization and denitrification processes have been developed to remove such materials.

A preferred way of treating the shale oil involves using a moving-bed reactor followed by a fractionation step to divide the wide-boiling-range crude oil produced from the shale oil into two separate fractions. The lower boiling fraction is hydrotreated for the removal of residual metals, sulfur, and nitrogen, whereas the heavier fraction is cracked in a second fixed-bed reactor normally operated under high-severity conditions.

The fluidized bed hydroretort process (Chapter 4) eliminates the retorting stage of conventional shale upgrading, by directly subjecting crushed oil shale to a hydroretorting treatment in an upflow, fluidized bed reactor such as that used for the hydrocracking of heavy petroleum residues. This process is a *single-stage retorting and upgrading* process. Therefore, the process involves (1) crushing oil shale; (2) mixing the crushed oil shale with a hydrocarbon liquid to provide a pumpable slurry; (3) introducing the slurry along with a hydrogen-containing gas into an upflow, fluidized bed reactor at a superficial fluid velocity sufficient to move the mixture upwardly through the reactor; (4) hydroretorting the oil shale; (5) removing the reaction mixture from the reactor; and (6) separating the reactor effluent into several components.

The mineral carbonate decomposition is minimized, as the process operating temperature is lower than that used in retorting. Therefore, the gaseous product of this process has a greater heating value than that of other conventional methods. In addition, owing to the exothermic nature of the hydroretorting reactions, less energy input is required per barrel of product obtained. Furthermore, there is practically no upper or lower limit on the grade of oil shale that can be treated.

Upgrading, or partial refining, to improve the properties of a crude shale oil may be carried out using different options (Lee et al., 2007; Speight, 2008; US DOE, 2004a, 2004b, 2004c). In conventional processing, Unocal catalytically hydrocracked raw kerogen oil. The process was both severe and costly, but resulted in a premium refinery feedstock. In the value-enhancement process, the raffinate, which has had its most problematic nitrogen removed, is hydrotreated under mild (approximately 300°C, 570°F) and low-cost conditions. This yields a refinery feedstock that is nearly as good as the Unocal feedstock. A comparison with Shell ICP oil shows the premium quality of this *in situ* oil, which is almost entirely atmospheric distillate. In general, oil from Green River oil shale has high hydrogen content and forms an excellent feedstock for manufacturing aviation turbine fuel and diesel fuel.

Gasoline from shale oil usually contains a high percentage of aromatic and naphthenic compounds that are not affected by the various treatment processes. The olefin content, although reduced in most cases by refining processes, will still be significant. It is assumed that diolefins and the higher unsaturated constituents will be removed from the gasoline product by appropriate treatment processes. The same should be true, although to a lesser extent, for nitrogen- and sulfur-containing constituents.

The sulfur content of raw shale oil gasoline may be rather high due to the high sulfur content of the shale oil itself and the frequently even distribution of the sulfur compounds in the various shale oil fractions. Not only the concentration but also the type of the sulfur compounds is of an importance when studying their effect on the gum formation tendency of the gasoline containing them.

Sulfides (R-S-R), disulfides (R-S-S-R), and mercaptans (R-SH) are, among the other sulfur compounds, and are the major contributors to gum formation in gasoline. Sweetening processes for converting mercaptans to disulfides should therefore not be used for shale oil gasoline; sulfur extraction processes are preferred.

Gasoline derived from shale oil contains varying amounts of oxygen compounds. The presence of oxygen in a product in which free radicals form easily is a cause for concern. Free hydroxy radicals are generated and the polymerization chain reaction is quickly brought to its propagation stage. Unless effective means are provided for the termination of the polymerization process, the propagation stage may well lead to an uncontrollable generation of oxygen-bearing free radicals leading to gum and other polymeric products.

Diesel fuel derived from oil shale is also subject to the degree of unsaturation, the effect of diolefins, the effect of aromatics, and to the effect of nitrogen and sulfur compounds. On the other hand, jet fuel produced from shale oil has to be subjected to suitable refining treatments and special processes. The resulting product must be identical in its properties to corresponding products obtained from conventional crude oil. This can be achieved by subjecting the shale oil product to a severe catalytic hydrogenation process with subsequent addition of additives to ensure resistance to oxidation.

If antioxidants are used for a temporary reduction of shale oil instability, they should be injected into the shale oil (or its products) as soon as possible after production. The antioxidant types and their concentrations should be determined for each particular case separately.

The antioxidants combine with the free radicals or supply available hydrogen atoms to mitigate the progress of the propagation and branching processes. When added to the freshly produced unstable product, the antioxidants may be able to fulfill this purpose. However, when added after some delay, that is, after the propagation and the branching processes have advanced beyond controllable limits, the antioxidants would not be able to prevent formation of degradation products.

Thus, shale oil is different from conventional crude oils, and several refining technologies have been developed to deal with this. The primary problems identified in the past were arsenic, nitrogen, and the waxy nature of the crude oil. Nitrogen and wax problems were solved using hydroprocessing approaches, essentially classical hydrocracking and the production of high-quality base stocks for lubricating oil manufacture, which require that waxy materials be removed or isomerized. However, the arsenic problem remains (DOE, 2004b).

In general, oil shale distillates have a much higher concentration of high boiling point compounds, which would favor production of middle distillates (such as diesel and jet fuels) rather than naphtha. Oil shale distillates also have a higher content of olefins, oxygen, and nitrogen than crude oil, as well as higher pour points and viscosities. Above-ground retorting processes tended to yield a lower API gravity oil than the in situ processes (a 25° API gravity shale oil was the highest API-gravity oil produced). Additional processing equivalent to hydrocracking would be required to convert oil shale distillates to a lighter-range hydrocarbon (gasoline). Removal of sulfur and nitrogen would, however, require hydrotreating.

Shale oil is often (and more correctly) referred to as synthetic crude oil and thus is closely associated with synthetic fuel production. However, the recovered hydrocarbons from oil shale are not yet economically competitive against petroleum crude. Furthermore, the value of hydrocarbons recovered from oil shale is diminished because of the presence of undesirable contaminants such as are sulfur-containing, nitrogen-containing, and metal-containing (organometallic) compounds, which cause detrimental effects to various catalysts used in the subsequent refining processes (Speight, 2007). These contaminants are also undesirable because of their disagreeable odor, corrosive characteristics, and combustion products that further cause environmental problems.

Accordingly, there is a great interest in developing more efficient methods for converting the heavier hydrocarbon fractions found in shale oil into lighter-molecular-weight hydrocarbons. The conventional processes include catalytic cracking, thermal cracking, and coking. It is known that heavier hydrocarbon fractions and refractory materials can be converted to lighter materials by hydrocracking. These processes are most commonly used on liquefied coals or heavy residual or distillate oils for the production of substantial yields of low-boiling saturated products. They are also used to some extent on intermediates that are used as domestic fuels, and on still heavier cuts that are used as lubricants. These destructive hydrogenation or hydrocracking processes may be operated on a strictly thermal basis or in the presence of a catalyst.

The thermal cracking process is directed toward the recovery of gaseous olefins as the primarily desired cracked product, in preference to naphtha-range liquids. By this process, it is claimed that at least 15–20% of the feed shale oil is converted to ethylene, which is the most common gaseous product. Most of the shale oil feedstock is converted to other gaseous and liquid products. Other important gaseous products are propylene, 1,3-butadiene, ethane, and butanes. Hydrogen is also recovered as a valuable nonhydrocarbon gaseous product. Liquid products can comprise 40–50% w/w or more of the total product. Recovered liquid products include benzene, toluene, xylene, gasoline-boiling-range liquids, and light and heavy oils.

Coke is a solid product of the process and is produced by polymerization of unsaturated materials. Coke is typically formed in an oxygen-deficient environment via dehydrogenation and aromatization. Most of it is removed from the process as a deposit on the entrained inert heat carrier solids.

Arsenic is removed from the oil by hydrotreating remains on the catalyst, generating a material that is a carcinogen, an acute poison, and a chronic poison. The catalyst must be removed and replaced when its capacity to hold arsenic is reached. Unocal found that its disposal options were limited.

The high pour point and the high viscosity of the raw shale oil and of the hydrotreated shale oil are a cause for concern. It appears that both the raw shale oil with its high nitrogen content, its high pour point, and high viscosity and the hydrotreated shale oil with its high pour point may not be suitable for undedicated (to such oils) pipelines. In the absence of dedicated pipelines, conversion to pipeline-acceptable products (such as gasoline, diesel fuel, and jet fuel) at or near the retorting site is one alternative. Another is to subject the raw shale to a coking operation, which lowers the pour point and, when followed by hydrotreating, gives a low-sulfur, low-nitrogen product with a pour point in the order of 7°C (45°F).

Upgrading crude shale oil at the mine site might consist of all of the above steps, although hydrogen-addition reactions generally predominate, and reactions to produce specialty chemicals are not likely to occur at all. Upgrading is typically directed only at the gaseous and liquid fractions of the retorting products and is rarely applied to the solid char that remains with the inorganic fraction of the oil shale, although coking of that solid fraction is possible. The most likely end products will be refinery feedstocks suitable for the production of middle distillates (kerosene, diesel fuel, jet fuel, and No. 2 fuel oil), although lighter-weight fuel components such as gasoline can also be produced. In general, hydrotreating followed by hydrocracking will produce jet fuel feedstocks, hydrotreating followed by fluid catalytic cracking is performed for production of gasoline feedstocks, and coking followed by hydrotreating is performed with the intention of producing diesel fuel feedstocks (Speight, 1997).

Similar to the preliminary steps taken at refineries, prior to or coincident with crude shale oil upgrading reactions, there are also activities to separate water from both the gas and the liquid fractions, to separate oily mists from the gaseous fraction, and to separate and further treat gases evolved during retorting to remove impurities and entrained solids and improve their combustion quality. Actions to remove heavy metals and inorganic impurities from crude shale oils also take place.

6.8 THE FUTURE

Projections that peak petroleum production may occur in the coming decades, along with increasing global demand, underscore the United States' dependence on imported petroleum. After Hurricanes Katrina and Rita, the spike in crude oil price and the temporary shutdown of some Gulf Coast refineries exacerbated that dependency. With imports making up 65% of the U.S. crude oil supply and the expectation that this percentage will rise, proponents of greater energy independence see the huge but undeveloped oil shale resources of the United States as one of several promising alternatives (Speight, 2007, 2011, 2013).

Oil shale is the least understood of the fossil fuel resources examined but new technologies, still in the research and development phase, have the potential to drastically alter the economics of its production. Despite a long history of activity in the oil shale industry, there is no large body of industrial knowledge based on successful operations from which to draw, so published costs for oil shale production have ranged from $10 to $95 per barrel.

Thus, it is not surprising that the failure of oil shale has been tied to the perennially lower price of crude oil, a much less risky conventional resource. Proponents of renewing commercial oil shale development might also decide whether other factors detract from the resource's potential. Refining industry profitability is overwhelmingly driven by light passenger vehicle demand for motor gasoline, and oil shale distillate does not make ideal feedstock for gasoline production. Policies that discourage the wider use of middle distillates as transportation fuels indirectly discourage oil shale development. Because the largest oil shale resources reside on federal lands, the federal government would have a direct interest and role in the development of this resource.

Opponents of federal subsidies for oil shale argue that the price and demand for crude oil should act as sufficient incentives to stimulate development. Projections of increased demand and peaking petroleum production in the coming decades tend to support the price-and-supply incentive argument in the long term.

Nevertheless, oil shale still has a future and remains a viable option for the production of liquid fuels. Many of the companies involved in earlier oil shale projects still hold their oil shale technology, and resource assets. The body of knowledge and understanding established by these past efforts provides the foundation for ongoing advances in shale oil production, mining, retorting, and processing technology, and supports the growing worldwide interest and activity in oil shale development. In fact, in many cases, the technologies developed to produce and process kerogen oil from shale have not been abandoned, but rather *mothballed* for adaptation and application at a future date when market demand increases and major capital investments for oil shale projects could be justified.

The fundamental problem with all oil shale technologies is the need to provide large amounts of heat energy to decompose the kerogen to liquid and gas products. More than 1 ton of shale must be heated to temperatures in the range of 850° and 1000°F (425–525°C) for each barrel of oil generated, and the heat supplied must be of relatively high quality to reach retorting temperature. Once the reaction is complete, recovering sensible heat from the hot rock is very desirable for optimum process economics. This leads to three areas where new technology could improve the economics of oil recovery: (1) recovering heat from the spent shale; (2) disposal of spent shale, especially if the shale is discharged at temperatures where the char can catch fire in the air; and (3) concurrent generation of large volumes of carbon dioxide.

Heat recovery from hot solids is generally not efficient, unless it is in the area of fluidized bed technology. However, to apply fluidized bed technology to oil shale would require grinding the shale to sizes less than about 1 mm, an energy-intensive task that would result in an expensive disposal problem. However, such fine particles might be used in a lower temperature process for sequestering carbon dioxide.

Disposal of spent shale is also a problem that must be solved in an economic fashion for the large-scale development of oil shale to proceed.

Retorted shale contains carbon as char, representing more than half of the original carbon content of the shale. The char is potentially pyrophoric and can burn if dumped into the open air while hot. The heating process results in a solid that occupies more volume than the fresh shale because of the problems of packing random particles. A shale oil industry producing 100 000 barrels per day, about the minimum for a world-scale operation, would process more than 100 000 tons of shale (density about 3 g/cc) and result in more than 35 m^3 of spent shale; this is equivalent to a block more than 100 feet on each side (assuming some effort at packing to conserve volume). Unocal's 25 000 bpd project of the 1980s filled an entire canyon with spent shale over several years of operation. Part of the spent shale could be returned to the mined-out areas for remediation, and some can potentially be used as feed for cement kilns.

In situ processes such as the Shell ICP process (Chapter 5) avoid the spent shale disposal problems because the spent shale remains where it is created (Fletcher, 2005a, 2005b, 2005c). In addition, ICP avoids carbon dioxide decomposition by operating at temperatures less than about 350°C (650°F). On the other hand, the spent shale will contain uncollected liquids that can leach into groundwater, and vapors produced during retorting can potentially escape to the aquifer. Shell has gone to great efforts to design barrier methods for isolating its retorts to avoid these problems (Mut, 2005). Control of in situ operation is a challenge that Shell claims to have solved in its work (Karanikas et al., 2005; Mut, 2005).

Shale (such as the Colorado shale) that contains a high proportion of dolomitic limestone (a mixture of calcium and magnesium carbonates) thermally deposes under retorting conditions and releases large volumes of carbon dioxide. This consumes energy and leads to the additional problem of sequestering the carbon dioxide to meet global climate change concerns.

In addition, there are also issues with the produced shale oil that also need resolution.

Unocal found that its shale oils contain arsenic. It developed a specialty hydrotreating catalyst and process, called SOAR for Shale Oil Arsenic Removal. This process was demonstrated successfully in the 1980s and is now owned by UOP as part of the hydroprocessing package purchased from Unocal in the early 1990s. Unocal also patented other arsenic removal methods.

Arsenic removed from the oil by hydrotreating remains on the catalyst, generating a material that is a carcinogen, an acute poison, and a chronic poison. The catalyst must be removed and replaced when its capacity to hold arsenic is reached. Unocal found that its disposal options were limited. Today, regulations require precautions to be taken when a reactor is opened to remove a catalyst.

Thus, several issues need to be resolved before an oil shale industry can be viable. These issues are not insurmountable but require a search for viable alternatives.

For example, one little explored alternative involves chemical treatment of shale to avoid the high-temperature process. The analogy with coal liquefaction here is striking: liquids can be generated from coal in two distinct ways: (1) by pyrolysis, creating a char coproduct or (2) by dissolving the coal in a solvent in the presence of hydrogen.

However, no similar *dissolution approach* to oil shale conversion is known because the chemistry of kerogen is markedly different from the chemistry of coal.

As a first step in developing a direct route, some attempts were made in the 1970s to isolate kerogen from the oil shale by dissolving away the minerals. Acid treatment to dissolve the mineral carbonate followed by fluoride treatment to remove the aluminosilicate minerals might be considered. Such a scheme will only work if the kerogen is not chemically bonded to the inorganic matrix. However, if the kerogen is bonded to the inorganic matrix, the bonding arrangement must be defined for the scheme to be successful.

Opportunities for circumventing the arsenic problem include development of an in-reactor process for regenerating the catalyst, collecting arsenic in a safe form away from the catalyst, and development of a catalyst or process where the removed arsenic exits the reactor in the gas or liquid phase to be scrubbed and confined elsewhere.

Shale oil produced by both aboveground and in situ techniques in the 1970s and 1980s were rich in organic nitrogen. Nitrogen compounds are catalyst poisons in many common refinery processes such as fluid catalytic cracking, hydrocracking, isomerization, naphtha reforming, and alkylation. The standard method for handling nitrogen poisoning is hydrodenitrogenation (HDN).

Hydrodenitrogenation is a well-established high-pressure technology using nickel molybdenum catalysts. It can consume prodigious amounts of hydrogen, typically made by steam reforming of natural gas, with carbon dioxide as a by-product.

Thus, after a decline of production since 1980 and the current scenarios that face a petroleum-based economy, the perspectives for oil shale can be viewed with a moderately positive outlook. This perspective is prompted by the rising demand for liquid fuels, the rising demand for electricity, as well as the change of price relationships between oil shale and conventional hydrocarbons.

Experience in Estonia, Brazil, China, Israel, Australia, and Germany has already demonstrated that fuels and a variety of other products can be produced from oil shale at reasonable, if not competitive, cost. New technologies can raise efficiencies and reduce air and water pollution to sustainable levels, and if innovative approaches are applied to waste remediation and carbon sequestration, oil shale technology can take on a whole new perspective.

In terms of innovative technologies, both conventional and in situ retorting processes result in inefficiencies that reduce the volume and quality of the produced shale oil. Depending on the efficiency of the process, a portion of the kerogen that does not yield liquid is either deposited as coke on the host mineral matter or is converted to hydrocarbon gases. For the purpose of producing shale oil, the optimal process is one that minimizes the regressive thermal and chemical reactions that form coke and hydrocarbon gases and maximizes the production of shale oil. Novel and advanced retorting and upgrading processes seek to modify the processing chemistry to improve recovery or create high-value by-products. Novel processes are being researched and tested in lab-scale environments. Some of these approaches include lower heating temperatures, higher heating rates, shorter residence times, introducing scavengers such as hydrogen (or hydrogen transfer/donor agents), and introducing solvents (Baldwin, 2002).

Finally, the development of western oil shale resources will require water for plant operations (Chapter 7), supporting infrastructure, and associated economic growth in the region. While some oil shale technologies may require reduced process water requirements, stable and secure sources of significant volumes of water may still be required for large-scale oil shale development. The largest demands for water are expected to be for land reclamation and to support the population and economic growth associated with oil shale activity.

Nevertheless, if a technology can be developed to economically recover oil from oil shale, the potential is enormous. If the kerogen could be converted to oil, the quantities would be far beyond all known conventional oil reserves. Unfortunately, the prospects for oil shale development are uncertain (Bartis et al., 2005). The estimated cost of surface retorting remains high and many consider it unwise to move toward near-term commercial efforts.

However, advances in thermally conductive in situ conversion may cause shale-derived oil to become competitive with current high grade crude oil. If this becomes the case, oil shale development could soon occupy a very prominent position in the national energy agenda. Only when it is clear that at least one major private firm is willing to devote, without appreciable government subsidy, technical, management, and financial resources to oil shale development, will government decision-makers address the policy issues related to oil shale development.

In 2005, Congress conducted hearings on oil shale to discuss opportunities for advancing technology that would facilitate *environmentally friendly* development of oil shale and oil sands resources (US Congress, 2005). The hearings also considered the legislative and administrative actions necessary to provide incentives for industry investment, as well as exploring concerns and experiences of other governments and organizations and the interests of industry. The Energy Policy Act of 2005 included provisions under Section 369 (Oil Shale, Tar Sands, and Other Strategic Unconventional Fuels) that direct the Secretary of the Interior to begin leasing oil shale tracts on public lands and to cooperate with the Secretary of Defense in developing a program to commercially develop oil shale, among other strategic unconventional fuels.

REFERENCES

Andrews, A., 2006. Oil Shale: History, Incentives, and Policy. Specialist, Industrial Engineering and Infrastructure Policy Resources, Science, and Industry Division. Congressional Research Service, the Library of Congress, Washington, DC.

Baldwin, R.M., 2002. Oil Shale: A Brief Technical Overview. Colorado School of Mines, Golden, CO, July.

Bartis, J.T., LaTourrette, T., Dixon, L., Peterson, D.J., Gary, C., 2005. Oil Shale Development in the United States: Prospects and Policy Issues. Prepared for the National Energy Technology of the United States Department of Energy. Rand Corporation, Santa Monica, CA.

Fletcher, S., 2005a. Industry, US government take new look at oil shale. Oil Gas J. April 11.

Fletcher, S., 2005b. Efforts to tap oil shale's potential yield mixed results. Oil Gas J. April 18, 2005.

Fletcher, S., 2005c. North American unconventional oil a potential energy bridge. Oil Gas J. April 11.

Guo, S., Qin, K., 1993. Nitro-containing compounds in Chinese light shale oil. Oil Shale 10 (2–3), 165–177.

Hunt, V.D., 1983. Synfuels Handbook. Industrial Press, New York. 1–216.

Karanikas, J.M., de Rouffignac, E.P., Vinegar, H.J., Wellington, S., 2005. In Situ Thermal Processing of an Oil Shale Formation While Inhibiting Coking. United States Patent 6,877,555, April 12.

Kundu, A., Dwivedi, N., Singh, A., Nigam, K.D.P., 2006. Hydrotreating catalysts and processes—current status and path forward. In: Lee, S. (Ed.), Encyclopedia of Chemical Processing, vol. 2. Taylor and Francis, Philadelphia, PA, pp. 1357–1366.

Lee, S., 1991. Oil Shale Technology. CRC-Taylor and Francis Group, Boca Raton, FL.

Lee, S., Speight, J.G., Loyalka, S.K., 2007. Handbook of Alternative Fuel Technologies. CRC-Taylor and Francis Group, Boca Raton, FL.

Mushrush, G.W., Speight, J.G., 1995. Petroleum Products: Instability and Incompatibility. Taylor and Francis, Washington, DC.

Mut, S., 2005. Testimony before the United States Senate Energy and Natural Resources Committee, Tuesday, April 12. http://energy.senate.gov/hearings/testimony.cfm?id=1445&wit_id=4139

Scouten, C., 1990. Part IV: Oil Shale. Chapters 25-31. In: Speight, J.G. (Ed.), Fuel Science and Technology Handbook. Marcel Dekker Inc., New York.

Speight, J.G., 2007. The Chemistry and Technology of Petroleum, fourth ed. CRC-Taylor and Francis Group, Boca Raton, FL.

Speight, J.G., 2008. Synthetic Fuels Handbook: Properties, Processes, and Performance. McGraw-Hill, New York.

Speight, J.G., 2011. An Introduction to Petroleum Technology, Economics, and Politics. Scrivener Publishing, Salem, MA.

Speight, J.G., 2013. The Chemistry and Technology of Coal, third ed. CRC-Taylor and Francis Group, Boca Raton, FL.

Tsai, T.C., Albright, L.F., 2006. Thermal Cracking of Hydrocarbons. In: Lee, S. (Ed.), Encyclopedia of Chemical Processing, vol. 5. Taylor and Francis, Philadelphia, PA, pp. 2975–2986.

US Congress, 2005. The Senate Energy and Natural Resources Committee, Oversight Hearing on Oil Shale Development Effort, April 12.

US DOE, 2004a. Strategic Significance of America's Oil Shale Reserves, I. Assessment of Strategic Issues, March. http://www.fe.doe.gov/programs/reserves/publications

US DOE, 2004b. Strategic Significance of America's Oil Shale Reserves, II. Oil Shale Resources, Technology, and Economics, March. http://www.fe.doe.gov/programs/reserves/publications

US DOE, 2004c. America's Oil Shale: A Roadmap for Federal Decision Making; USDOE Office of US Naval Petroleum and Oil Shale Reserves. http://www.fe.doe.gov/programs/reserves/publications

CHAPTER 7

Environmental Aspects

7.1 INTRODUCTION

Oil shale will, unfortunately, contribute to environmental pollution, including phenomena such as acid rain, the greenhouse effect, and allegedly global warming (global climate change) (Speight, 2008). Whatever the effects, the risks associated with the oil shale fuel cycle can be minimized by introducing relevant technologies for the protection of the environment.

Until the 1960s, a legacy of unchecked fossil fuel use was environmental damage, which spurred the United States Congress to introduce federal regulations to limit impacts to the environment. This legacy includes physical disturbances to the landscape, subsidence and settlement above abandoned underground mines, flooding and increased sedimentation, polluted groundwater and surface water, unstable slopes, and public safety and land disturbance issues. In countries where such regulations do not exist, these issues are a major concern for the development of oil shale resources.

Oil shale itself is considered to be harmless, and presents no risk when it remains in situ where it was formed and deposited millions of years ago. When involved recovery and conversion activities, however, its environmental impacts are deleterious if oil shale is utilized in the wrong place at the wrong time in wrong amounts. Thus, concerns about the impacts of oil shale on the environment and human health are not new.

Oil shale in the ground does not generally pose an environmental threat, although mineralogy can influence groundwater properties.

However, oil shale production and use has diverse impacts on the surrounding earth and atmosphere, generating various pollutants. Irrespective of how it is extracted in mines and used in industry, oil shale produces three distinct types of pollutants: (1) gaseous, (2) liquid, and (3) solid pollutants. They generally demand quite unique preventive or ameliorative measures. In this context, other impacts like noise, subsidence, and waste disposal should also be classified as pollutants arising from oil shale use. Numerous methods have been devised to keep environmental standards at threshold limits and thus minimize pollution damage while improving worker productivity, oil shale quality, and accident prevention schemes.

Oil shale mining, like coal mining, remains a dangerous occupation (Speight, 2013). Mine atmospheres are hazardous in terms of the health and safety of miners and, although technological advancements and legislation have led to improvements in the environmental and safety aspects of mines, pollutants are still produced at significant levels from oil shale excavation in both surface and underground mines.

The production of airborne dust particles is a major problem in underground mines where dust explosions, due to released gases, are the main concern. In addition, dust is not solely a localized problem – fine particles can be transported far from the source to contaminate other areas.

Dust is a by-product of blasting and (in open-pit mines) earthmoving operations. The degree of risk depends on the physical size of the particulate matter, the humidity of surrounding air, and the velocity and direction of prevailing winds. Particle size and duration of exposure determine how far dust and droplets penetrate the respiratory tract. Inhaled fine dust remains in the alveoli, but all types of larger particles are removed by the filtering mechanism of the respiratory system.

In surface working sites water spraying, site selection, and screening from winds are advantageous in removing dust, whereas in underground working sites suppression by water, dilution and dispersion by ventilation, and removal by filtration or electrostatic precipitation are advisable.

Significant volumes of earth must be displaced to mine oil shale, and the resulting rock waste can disrupt the environment. Furthermore, waste material from derived from the sinking of shafts, roadways, and ventilation tunnels, and extraction of the oil shale sediment during mining operations.

The enormous volumes of waste material that are the by-products of both underground and surface excavation operations are a major pollution issue, and disposal of these solid wastes is the most controversial aspect of oil shale mining. The primary environmental damage of waste piles (mine tips) are not only noise and dust from moving vehicles but also groundwater contamination, leaching of toxic and acid pollutants, and loss of usable land. Reclamation restores the land, but reduction in the fertility of the soil and diminished ecological habitat are slow to recover. In fact, old mine tips, inherited from unregulated past practices of dumping mine waste, are a danger to local drainage systems and may have toxic effects on human health.

Once a mine is worked out and oil shale recovery operations cease, mined out areas should be converted into productive agricultural land or restored to their former natural beauty. Potential recovery work comprises topsoil and subsoil replacement, compaction, regrading, revegetation of the land, and chemical treatment and management of contaminated water resources. Recultivation necessitates the enrichment of the soil for seeding and planting. Generally, land restoration reduces the potential for land destruction and pollution hazards.

Furthermore, all approaches to oil shale exploitation—aboveground retorting, in situ operations, and modified in situ operations—must address substantial environmental and health and safety concerns. The regulatory structure is already in place, and the projects have to comply with the Clean Water Act, the Clean Air Act, and numerous other federal and state regulations that govern industrial operations. Some research, development, and testing may be required to ensure compliance with such regulations, because no large-scale industry currently exists to provide a database.

The most serious environmental concerns of oil shale mining are concerns associated with the management and disposal of solid waste, especially the rock that remains after shale oil has been extracted.

Commercial-scale aboveground retorting operations will generate huge quantities of retorted and spent shale, which will contain soluble salts, organic compounds, and trace concentrations of numerous heavy metals. Regardless of where they are deposited, oil-shale-processing wastes must be protected from leaching by snowmelt, rainfall, and groundwater, because leached salts and toxins can contaminate both aquifers and surface streams. Toxics, including carcinogens, mutagens, priority pollutants, and other hazardous substances, have been reported in various types of oil-shale-processing wastes (Kahn, 1979).

Any toxic substance present in wastewater streams must be removed along with trace organic or inorganic substances. It is not expected that thermal oxidation, which is often employed to destroy hazardous organic compounds, will be required for treating such wastewater streams, although it may be considered for treating concentrates or sludges. However, the presence of toxic substances may interfere with biological oxidation processes used for the removal of bulk organic matter. If this is a problem, the substances can be removed by any of several conventional pretreatment steps.

In the case of in situ and modified in situ operations, the retorted shale will be left underground, that is, out of sight and out of reach but potentially exposed to groundwater infiltration and leaching. If infiltration occurs, it could be very difficult to confine the contamination because there will be little access to the affected areas.

Air quality will also be threatened by fugitive dust, acidic gases, and combustion products from retorts, heaters, and electrical generators. This concern also affects all approaches to shale oil extraction, as does the potential for surface subsidence.

The interest in retorting oil from oil shale in order to produce a competitively priced synthetic fuel has intensified since the oil embargo of the 1970s. Commercial interest was very high in the 1970s and 1980s, but decreased substantially in the 1990s due to the fluctuation (words to low end of the scale) of oil prices. However, interest in oil shale for producing clean liquid fuel is being revitalized in the twenty-first century – mainly triggered by the varying (upwards) petroleum prices as well as the shortage of oil in the global market.

With this renewed interest and stricter environmental regulations comes the higher interest in environmental issues that always accompany the development of any fossil fuel resource.

The technology that is used to produce shale oil is dependent on the depth, thickness, richness, and accessibility of a deposit. Deeper and thicker beds will likely be worked in situ. A combination of approaches will likely be used in the western U.S. basins. Various land impacts are associated with oil shale processing. Open-pit (surface) mining involves significant surface disturbance and can impact surface water runoff patterns and subsurface water quality. Evidence from coal mining and other mining industries has demonstrated that impacted lands can be very effectively reclaimed with minimal long-term effects (Speight, 2013).

Air and water quality, topography, wildlife, and the health and safety of workers will be affected by the development of an oil shale industry. Many effects will be similar to those caused by any type of mineral development, but the scale of operations, their concentration in a relatively small geographic area, and the nature of generated wastes present some unique challenges. The environmental impacts of this industry will be regulated by state and federal laws.

The potential leaching of waste disposal areas and in situ retorts once the oil shale plants are abandoned is a major concern. If this occurs, leachates may degrade the water quality in any nearby water system. Such *nonpoint* wastewater discharges are neither well understood nor well regulated, although the Clean Water Act provides a regulatory framework. Techniques for preventing leaching need to be demonstrated on a commercial scale. It is necessary to test a variety of development technologies to ensure adequate control of a large industry.

Developers of oil shale leases would do well to look at Estonia, to observe, and learn from the development of Estonian oil shale tracts. Oil shale developers would also do well to look to Canada, especially Alberta, where the development of tar sand (oil sand) leases has been commercialized since 1967 and can also give some pointers on protecting the environment.

This chapter focuses on the environmental and human health issues related to the development of oil shale resources. Emphasis is placed on those issues that are related to the petrographic, chemical, and mineralogical composition of oil shale, recognizing that a balance must be struck between industrial development and energy requirements in order to build self-contained national economies.

7.2 AIR QUALITY IMPACT

The Clean Air Act is the only existing environmental law that can prevent the creation of a large industry. The procedures for obtaining environmental permits can take several years, which should not preclude the establishment of an individual project, and they must be factored into overall project costs.

Oil shale is a carbonate rock that when heated to 450–500°C creates kerogen oil and hydrocarbon gases along with a slate of other gases, which may include (1) oxides of sulfur and nitrogen, (2) carbon dioxide, (3) particulate matter, and (4) water vapor. Commercially available stack gas cleanup technologies that are currently in use in electric power generation and petroleum-refining facilities have improved over the years and should be effective in controlling emissions of oxides and particulates from oil shale projects.

7.2.1 Dust Emissions and Particulate Matter

The production of airborne dust particles is a major problem in underground mines, where dust explosions, with or without the release of gases, are the main concern. In addition, dust is not solely a localized problem since fine particles can be transported to contaminate areas far from their source.

Dust is a by-product of blasting and (in open-pit mines) earthmoving operations. Its degree of risk depends on the physical size of particulate matter, humidity of surrounding air, and velocity and direction of prevailing winds. Particle size and duration of exposure determine how far dust and droplets penetrate the respiratory tract. Inhaled fine dust remains in the alveoli, but all types of larger particles are removed by the filtering mechanism of the respiratory system. The constituents of particulate matter (*particulates*) having a diameter less than 10 microns (particularly those in the range of 0.25–7 microns) may result in respiratory diseases such as chronic bronchitis and pneumoconiosis. If the dust contains silica particles, diseases such as silicosis (progressive nodular fibrosis) become a major threat to human health.

Operations such as crushing, sizing, transfer conveying, vehicular traffic, and wind erosion are typical sources of fugitive dust. Control of airborne particulate matter (PM) could pose a challenge. Compliance with regulations regarding particulate matter control must be factored in to operational planning.

In surface working sites, water spraying, site selection, and screening from winds are advantageous in removing dust, whereas in underground working sites suppression by water, dilution and dispersion by ventilation, and removal by filtration or electrostatic precipitation are advisable.

In addition to the hazards to health, dust in a mine can contribute to serious explosion hazards. Provision of efficient ventilation of the mine and restrictions to electrical or thermal sources of ignition are required by regulation.

The main sources of dust emissions are crushers, breakers, dedusters, dry screens, and transfer points between units. The size of oil shale particles that become airborne generally falls into the range of 1–100 micron (i.e., 1–100 × 10^{-6} m), and particles larger than this range usually deposit close to the point of origin. It is the usual practice to extract dust clouds from various sources through ducts that exhaust to a central air-cleaning unit, although some equipment may include integral dust suppression devices. Dust collection equipment includes (1) dry units such as cyclones, dynamic collectors, and baghouses (fabric filters); and (2) wet units such as dynamic, impingement, or centrifugal devices, or gravity, disintegrator, or venturi scrubber systems.

Other sources of dust are windblown losses from storage piles and from railcars during loading and transportation. Various chemical treatments are available for stockpile sealing and are used for spraying the tops of railcars. However, active stockpiles present an almost insurmountable problem.

7.2.2 Mine Waste Disposal

Similar to coal mining (Speight, 2013), significant volumes of earth must be displaced to mine oil shale, and the resulting rock waste can disrupt the environment.

Waste material from deep mines derives from the sinking of shafts, roadways, and ventilation tunnels, and extraction of the oil shale. This is then hauled to the surface and dumped locally. However, the use of modern mechanical techniques, particularly the introduction of rock-cutting machines, has dramatically increased the proportion of waste material.

The enormous volumes of waste material that are the by-products of both underground and surface excavation operations are a major pollution issue, and disposal of these solid wastes is the most controversial aspect of oil shale mining. The primary environmental damage of waste piles (mine tips) are not only noise and dust from moving vehicles but also groundwater contamination, leaching of toxic and acid pollutants, and loss of usable land. Reclamation will restore the land, but reduction in the fertility of the soil and diminished ecological habitat are slow to recover. In fact, old coal mine tips, inherited from unregulated past practices of dumping mine waste, are a potential danger to local drainage systems and may have toxic effects on human health.

Another issue related to waste disposal is as follows: the oxidation of pyrite produces acidic compounds, which, with other toxic materials, can leach into the local water supply. Although pyrite is not as plentiful in oil shale as it is in coal, its presence is still dangerous. Simultaneously, heat produced from such chemical reactions can lead to the spontaneous combustion of organic particles in waste tips. The potential hazards of this process in spoil heaps can be substantially reduced by controlled tipping, careful site selection, as well as compaction of waste.

Surface mining has a greater adverse effect on the surroundings than underground mining operations. For example, in strip mining operations, the overburden is removed and the volume of the mined overburden may be many times the volume of oil shale produced. In addition, the mining operations (which use power shovels, draglines, and bucket-chain/bucket-wheel excavators) alter the topography of the surface, destroy all the original vegetation in the area, and often lead to contamination of surface water and groundwater courses. Nevertheless, surface mining is not considered to be a major contributor to air pollution.

Rock waste dumped indiscriminately during surface and underground mining processes weathers rapidly and has the potential for producing *acid drainage*, which is a source of oxygen- and sulfur-containing compounds that combine with water to form acidic species. In the past, mine overburden or mine tipple was usually dumped in low-lying areas, often filling wetlands or other sources of water. This resulted in the dissolution of heavy metals, which seeped into both groundwater and surface water causing disruption of marine habitats and deterioration of drinking water sources. In addition, pyrite (FeS_2) can form sulfuric acid (H_2SO_4) and iron hydroxide [$Fe(OH)_2$] when exposed to air and water. When rainwater washes over these rocks, the runoff can become acidified, affecting local soil environments, rivers, and streams (*acid mine drainage*) (Berkowitz, 1985; Speight, 2013). The extent and toxicity of such waste streams depend on oil shale characteristics, local rainfall patterns, local topography, and site drainage features. Leaching of such waste could lead to an unacceptable level of contamination of surface water and groundwater.

Area mining occurs on level ground, where workers use excavation equipment to dig a series of long parallel strips, or *cuts*, into the earth. The overburden is cleared from each cut, and the material (known as *spoil*) is stacked alongside the long trench. After the exposed mineral is retrieved from the cuts, workers dump the spoil back into the trench to help reclaim the mined area.

Once the mine is worked out and recovery operations cease, mined-out areas are to be converted into productive agricultural land or restored to their former natural beauty. Potential recovery work comprises topsoil and subsoil replacement, compaction, regrading, revegetation of the land, and chemical treatment and management of contaminated water resources. Recultivation necessitates the enrichment of soil for seeding and planting. Generally, land restoration reduces the potential for land destruction and pollution hazards.

7.2.3 Hazardous Air Pollutants

Gaseous emissions such as hydrogen sulfide (H_2S), ammonia (NH_3), carbon monoxide (CO), sulfur dioxide (SO_2), nitrogen oxides (NO_x), and trace metals species are sources of air pollution. Such emissions are at least conceivable in oil-shale-processing operations. However, the level of severity in this case is far less than that of the processing operations of other fossil fuels. The very same argument can be made for the emission of carbon dioxide, which is a major greenhouse gas.

The lifetime of these pollutant species in the atmosphere is relatively short, and if they are distributed evenly their harmful effects will be minimal. Unfortunately, these human-made effluents are usually concentrated in localized areas and their dispersion is limited by both meteorological and topographical factors. Furthermore, synergistic effects mean that the pollutants interact with each other in the presence of sunlight, carbon monoxide, nitrogen oxides, and unburned hydrocarbons, leading to photochemical smogs; when sulfur dioxide concentrations become appreciable, sulfur-oxide-based smog is formed.

It is worth remembering that even the nontoxic, but non-life-supporting and suffocating, carbon dioxide may have an important effect on the environment. The surface of the Earth emits infrared radiation with a peak of energy distribution in the region where carbon dioxide is a strong absorber. This results in a situation in which this infrared radiation is trapped by the atmosphere and the temperature of the Earth's surface is increased. As a result of the combustion of fossil fuels, the concentration of carbon dioxide in the atmosphere is continuously increasing. Although many factors are involved, it does seem that an increase in the carbon dioxide concentration in the atmosphere would result in a rise in temperature at the surface of the Earth, which could cause an appreciable reduction in polar ice caps; this in turn would result in further heating of the Earth's surface.

Acid gases (SO_x and NO_x) emitted into the atmosphere during shale processing provide components essential to the formation of acid rain. Sulfur is present in oil shale in both organic and inorganic compounds. On processing, most of the sulfur is converted to sulfur dioxide with a small proportion remaining in the ash as sulfite:

$$S + O_2 \rightarrow SO_2$$

In the presence of excess air, some sulfur trioxide is also formed:

$$2SO_2 + O_2 \rightarrow 2SO_3$$

$$SO_3 + H_2O \rightarrow H_2SO_4$$

Even a small amount of sulfur trioxide can have adverse effects, as it brings about the condensation of sulfuric acid and causes severe corrosion.

Also, the nitrogen inherent in kerogen can be converted to nitrogen oxides during processing, which also produces acidic products and thereby contributes to the formation of acid rain:

$$2N + O_2 \rightarrow 2NO$$

$$NO + H_2O \rightarrow HNO_2$$

Nitrous acid

$$2NO + O_2 \rightarrow 2NO_2$$

$$NO_2 + H_2O \rightarrow HNO_3$$

Nitric acid

Most stack-gas-scrubbing processes are designed for sulfur dioxide removal; nitrogen oxides are controlled as far as possible by modification of combustion design and regulation of flame temperature

(Speight, 2007, 2013). However, processes that remove sulfur dioxide usually do remove some nitrogen oxides. PM can be removed efficiently by using commercially well-established electrostatic precipitators.

7.3 WATER QUALITY IMPACT

7.3.1 Water Quality

Suspended solids will occur primarily in water from the dust-control systems used in shale mining and crushing operations. Mine drainage water will also contain suspended solids, as will a retort condensate stream that picks up fine shale particles as it trickles down through the broken shale. In aboveground retorts, some fine shale may be entrained in the retort gas and captured in the gas condensate, but levels should be low, thus should not be a problem to treat. Cooling water will pick up dust from the atmosphere, particularly if the cooling tower is near a shale crushing operation or a shale disposal site. Precipitated salts and biological matter may also be present in the cooling tower blowdown.

Dissolved inorganics will be found in mine drainage water and retort condensates because these streams leach sodium, potassium, sulfate, bicarbonate, chloride, calcium, and magnesium ions from shale when they come in contact with it. In addition, some inorganic volatiles may be captured from the gas phase in the retort. Of the heavy metals present in raw oil shale, cadmium and mercury (probably as their respective sulfides) are expected to be present in the gas condensate in low concentrations.

Dissolved organics arise largely from the organic compounds in raw oil shale. They may be altered during pyrolysis and may end up in the retort, gas, or hydrotreater condensates. The types of organics in each condensate will probably depend on the volatility and volubility of the organics and the temperature at which the wastewater is condensed.

Water-soluble phenols accumulate in water layers of retorting-unit condensation systems, and additional quantities are obtained by washing the shale oil fractions with water. Phenols are removed from these phenol waters—phenols constitute up to 2% of the oil formed during oil shale retorting (one-third is dissolved in tar water, and two-thirds are obtained by supplementary washing of shale oil).

Toxic minerals and substances that are exposed during the removal of the overburden include acidic materials, highly alkaline materials, and dilute concentrations of heavy metals. These materials can have an adverse effect on indigenous wildlife by creating a hostile environment (often by poisoning the waterways) and, in some cases, causing the destruction of species. Thus, mine design should include plans to accommodate the potentially harmful substances generated by the weathering of spoil piles.

Being predominantly centered in the western United States, oil shale development will most likely take place in areas of historic interest, and this may well be the case in other countries. Artifacts on historic sites will be destroyed unless they have been systematically investigated before being disturbed by mining. A mining plan should include the provision for systematic archaeological studies of the area to be mined. Such studies by the mining industry often benefit the community and create an appreciation for historic values.

Finally, aesthetic rehabilitation of mined lands should be, and is being, done so that such lands are aesthetically more pleasing after mining than before. Removal of the overburden, however, is disruptive to the landscape and aesthetically repugnant for a temporary period until the land is restored. From the community and regulatory viewpoints, it is beneficial if a mining company includes some consideration of aesthetic values in the mining plan.

7.3.2 Water Requirements

A rate-limiting factor in developing the oil shale resources of the western United States is not only the effect of water quality but also the availability of water, which may not be a problem in the eastern regions of the United States. Water from the Colorado River could be made available for depletion by oil shale. An important factor that must be taken into consideration in any water-use plan is the potential salt loading of the Colorado River. With oil shale development near the river, its average annual salinity is anticipated to increase, unless some prevention or treatment is implemented. The economic damages associated with these higher salinity levels could be significant and have been the subject of extensive economic studies.

The water required for oil shale retorting is estimated to be one to three barrels of water per barrel of shale oil. Still, some processes may be net producers of water. For an oil shale industry producing 2.5 million barrels per day, this equates to between 105 and 315 million gallons of water per day. An oil shale industry producing 2.5 million barrels per day would require 0.18–0.42 million acre feet of water per year, depending on the location and the processes used.

In the western United States, water will be drawn from local and regional sources. The major water source is the Colorado River Basin, which includes the Colorado, Green, and White rivers. The Colorado River flows between 10 and 22 million acre feet per year. Water may also be purchased from other existing reservoirs. In addition, transfers may be possible from other water basins, including the Upper Missouri Basin. Another water source will come from the western oil shale itself, which has a high water content.

Oil shale typically holds 2–5 gallons of water per ton, although some oil shale can contain as much as 30–40 gallons of water per ton. Much of this connate water can be recovered during processing and used to support mining, disposal, or reclamation operations. Although produced water will contain organic and inorganic substances, its impurities can be removed using conventional water treatment technologies. Recycling and reuse of process water will help to reduce water requirements (US DOE, 2006). Produced water from other conventional and unconventional oil and gas operations may also provide a water source.

Development of oil shale resources in the western Unted States will require significant quantities of water for mining and plant operations, reclamation, supporting infrastructure, and associated economic growth. Initial process water requirement estimates of 2.1–5 barrels of water per barrel of oil, first developed in the 1970s, have gone down. More current estimates based on updated oil shale industry water budgets suggest that requirements for new retorting methods will be one to three barrels of water per barrel of oil (US OTA, 1980). Some processes may be net producers of water.

For an oil shale industry producing 2 500 000 barrels per day, this equates to between 105 and 315 million gallons of water per day. These numbers include water requirements for power generation for in situ heating processes, retorting, refining, reclamation, dust control, and on-site worker demands. Municipal and other water requirements related to population growth associated with industry development will require an additional 58 million gallons per day.

An oil shale industry producing 2 500 000 barrels per day would require 0.18–0.42 million acre feet of water per year, depending on the location and the processes used. Water supply issues are less critical for eastern oil shales where the supply is ample.

In the western United States, water will be drawn from local and regional sources. The major water source is the Colorado River Basin, which includes the Colorado, Green, and White rivers. The Colorado flows between 10 and 22 million acre feet per year. Water may also be purchased from other existing reservoirs. Transfers may be possible from other water basins, including the Upper Missouri.

Western oil shale has a high water content. Some oil shale contains 30–40 gallons per ton of shale. More typically it holds 2–5 gallons of water per ton. Much of this water can be recovered during processing and used to support operations. Produced water will contain organic and inorganic substances that can be removed using conventional filtering technologies. Recycling and reuse of process water will help to reduce water requirements.

Water in the western United States is treated much the same as other commodities—it can be bought and sold in a competitive market. Interstate "compacts" control the amount of river water each state is entitled to use. They allocate 5.3–5.9 million acre feet to the states. States are expected to use about 4.8 million acre feet of their allocations by 2020. If all industry water is withdrawn from the river, oil shale development will increase withdrawals by 0.18–0.42 million acre feet per year. Use of connate water and water reuse can reduce this volume. A system of rights and seniority has been established that allocates expected resources. Many private companies previously engaged in oil shale development retain very senior rights, which they had obtained during the 1970s. Because federal lands and prospective future leases will not come with water rights, some lessees may need to negotiate water purchases to advance their respective oil shale projects.

Initial estimates indicate that enough water will be available to support oil shale industry development in the western states. However, variability of supply during low-flow years may cause conflicts among water users.

As the industry grows, additional water resources for human consumption and for oil shale processes will likely be required. The water consumption growth will slow as oil shale technologies become more efficient. For a mature industry, substantial water storage and water transfers may be required over time. The overall allocation of water today is governed by the Colorado River Compact, originally agreed to on November 24, 1922. Currently, there is a mix of both absolute and conditional water rights in existence.

Absolute rights to water use are those that have been decreed by the state water court as available for use. Conditional rights to water use are rights that have not been through the Court process and therefore have not been decreed. They cannot be used until a decree has been granted and the rights have been determined to be absolute. Conditional rights only preserve a holder's seniority in accordance with the doctrine of first in time, first in right. In addition, conditional rights must undergo a diligence test every 6 years to preserve the conditional right. An absolute right to water use is still subject to being curtailed (a call) in the event that the water balance is insufficient for all rights and a senior right holder is being injured.

To help ensure supply, it is customary to file an augmentation plan, which may consist of a plan for reservoir storage and release or purchase of senior rights that can be provided to a senior right holder.

An agreement (signed in October 2003) between the State of California and the Upper Basin States returns about 0.8 million acre feet per year to the Upper Basin States that were being overused by the State of California. This increment of 0.8 million acre feet per year can help support an oil shale industry, if the water is largely allocated to this use (US DOE, 2004a, 2004b, 2004c).

Disposal of excess water is an integral and essential phase of mining operations, because pollutants in waterways result in the reduction of oxygen content of water leading to the destruction of aquatic life. In modern mining practices for controlling water flow, sumps and pools are provided where drainage water accumulates and where suspended shale and clay particles can ultimately be removed. Sedimentation, with or without the use of flocculants, is used to process water in lagoons. Occasionally, if geological conditions are favorable, an adequate thickness of strata is left as a wall or pillar between the aquifer and mine workings to prevent water flow into mines.

Remedies are mainly aimed at restricting water flows seeping through the porous structure, boreholes, and fractures in the water-bearing strata.

This can be brought about by sealing boreholes, grouting fracture zones, restricting the flow of free oxygen into mines, cave-ins to fill the voids, and diverting groundwater courses.

7.4 LAND QUALITY IMPACT

7.4.1 Subsidence

Subsidence is a costly economic impact of underground mining because it creates horizontal and vertical displacement of the surface, which generally causes structural damage to buildings, roads, and railroads, as well as pipeline ruptures.

Factors contributing to ground movements and ultimate surface damage are the thickness, dip and depth of the oil shale seam, angle of draw, the nature and thickness of the overburden, and the amount of support left in the *goaf* (*goff* or *gob*—a mine from which the mineral has been partially or wholly removed). In addition, seepage of methane through cracks into houses results in accumulation of methane and gas explosions causing excessive damage to property.

Subsidence is less likely to occur if the dimensions of the working areas are limited and permanent support practices are applied in the goaf. Surface destruction may be scaled down by backfilling the abandoned workings with stowing material or solid mine waste. However, such preventive methods to reduce ground movement are only one aspect of the problem since it must be achieved without excess loss of oil shale in the pillars and with minimum interference to normal mining operations.

Underground mine planning and design has as its goal an integrated mine systems design, whereby a mineral is extracted and prepared at a desired market specification and a minimum unit cost within acceptable social, legal, and regulatory constraints. A large number of individual engineering disciplines contribute to the mine planning and design process, making it a multidisciplinary activity. Given the complexity of the mining system, planning ensures the correct selection and coordinated operation of all subsystems, whereas design applies to the traditionally held engineering design of subsystems.

Planning must account for both environmental protection, beginning as early as initial exploration, and reclamation. It is critical that planning alleviates or mitigates potential impacts of mining for two key reasons: (1) The cost of environmental protection is minimized by incorporating it into the initial design, rather than performing remedial measures to compensate for design deficiencies, and (2) negative publicity or poor public relations may have severe economic consequences. From the start of the planning process, adequate consideration must be given to regulatory affairs. The cost of compliance may be significantly reduced when it is taken into account in the design or planning process, in a proactive manner, rather than being addressed on an ad hoc basis as problems develop or when enforcement actions occur.

From the beginning of the mine design planning stage, data gathering, permitting, and environmental considerations are important, although benefits in a strictly economic sense may be intangible. From exploration, where core holes must be sealed and the site reclaimed, through plan development, the impacts on the environment must be considered. These impacts include aesthetics, noise, air quality (dust and pollutants), vibration, water discharge and runoff, subsidence, and process wastes; sources include the underground and surface mine infrastructure, mineral processing plants, access or haul roads, and remote facilities. If mining will cause quality deterioration of either surface water or groundwater, remedial and treatment measures must be developed to meet discharge standards. The mine plan must include all the technical measures necessary to handle all the environmental problems from initial data gathering to mine closure and reclamation of the disturbed surface area.

Reclamation plans include many of the following concerns: drainage control, preservation of topsoil, segregation of waste material, erosion and sediment control, solid waste disposal, control of fugitive dust, regrading, and restoration of waste and mine areas. The plan must also consider the effects of mine subsidence, vibration (induced by mining, processing, transport, or subsidence), and impact on surface water and groundwater. These environmental items often dictate the economics of a planned mining operation and determine its viability.

The environmental aspects of underground mining are different from those encountered in surface mining operations. They have been considered to be different, even to the point of being considered (erroneously) to be of lesser importance. But it cannot be denied that underground mining can disturb aquifers either through the construction of a shaft or as a result of other influences such as formation disturbance due to subsidence. Indeed, it was only during the last three decades that subsidence was elevated to the role of a major environmental issue.

In addition, the transportation of spoil or tipple to the surface from an underground mine where it is then deposited in piles or rows offers a new environmental hazard. The potential for leaching materials from spoil/tipple offers a mode of environmental contamination that is not often recognized and that is of at least equal importance to the potential contamination from surface mining operations. Similarly, the discharge of gaseous and liquid effluents from underground mining operations should be of at least equal concern to those arising from surface mining operations. Thus, the environmental issues related to oil shale mining, be it underground or surface mining, are multifaceted.

Whether the objective is underground mining or surface mining, preparation of the site or stripping of the overburden from the oil shale formation is a trauma for the environment. Vegetation is removed, flora and microorganisms are disturbed and/or destroyed, soil and subsoil are removed, underlying strata are ruptured and displaced, hydrological systems may also suffer, and the surface is exposed to weathering (i.e., oxidation, which can result in the chemical alteration of mineral components); other general topographic changes also take place. For example, there is the ever-present danger of waterway contamination during (and after) mining. Materials in the overburden (such as heavy metals and/or minerals) can be leached from the overburden, usually after erosion by rain or by run-off water from snowfall, into a nearby waterways, and due caution must be used in such enterprises. However, from a more positive viewpoint, mining can also be beneficial to the local hydrology of an area; spoil piles may act as "sponges" that absorb large quantities of water for future plant growth and use, retard and diminish runoff, and provide aquifers for supplying flow to nearby streams on a steady basis.

Mining can and does cause some long-lasting wildlife habitat impairment or changes of habitat. A few wildlife species may be unable to adjust to these changes, and they do not return to the restored lands but live elsewhere in the neighborhood. But it is more than likely that, given sufficient time after the restoration of mined lands, the long-term impact of mining is favorable for wildlife. Long-term impacts of surface mining on wildlife can be minimized by careful consideration of wildlife presence when making plans for mining.

Toxic minerals and substances that are exposed during the removal of the overburden include acidic materials, highly alkaline materials, and dilute concentrations of heavy metals. These materials can have an adverse effect on indigenous wildlife by creating a hostile environment (often through poisoning the waterways) for them and, in some cases, causing the destruction of species. Thus, mine design should include plans to accommodate potentially harmful substances, which are generated by the weathering of spoil piles.

Finally, aesthetic rehabilitation of mined lands should be, and is being, done so that a land is aesthetically more pleasing after mining than before. The removal of the overburden, however, is disruptive to the landscape and aesthetically repugnant for a temporary period until the land is restored. From the community and regulatory viewpoints, it is beneficial if a mining company includes consideration of aesthetic values in the mining plan.

7.4.2 Noise, Vibration, and Visibility

Noise from underground mining operations does not have a significant environmental impact on a population, but it can be hazardous to miners working in the underground caverns. However, residential communities in the locality of open-pit mines often find the quality of life diminished due to the effects of noise from mining operations, vibration from blasting, and continuous heavy traffic. Furthermore, clouds of dust may reduce visibility and increase haze, producing severe distress and annoyance.

These adverse conditions can be improved to a certain degree by limiting excavation to one section of the mining area at any one time, as well as siting of boundaries and screening banks, erection of barriers, and use of low-noise machinery fitted with effective exhaust silencers. The implementation of environmentally friendly techniques has served to establish better public relations between mining companies and the local populace.

7.4.3 Reclamation

Surface and underground mining of oil shale can disrupt the premining environment. Vegetative cover must often be removed. Geological and soil profiles can be overturned so that rich soils are now buried and leachable, lifeless rock is on top. After mining is complete, reclamation of the land is necessary.

Thus, reclamation, as it applies to mining activities, means rehabilitating the land to a condition where it can support at least the same land uses as it supported prior to the mining activity. In some countries, reclamation may not involve restoring the land to its exact condition prior to mining but only to a condition where it can support the same or other desired productive land use. This can involve reestablishment of the biological community (flora and fauna), or alternate biological community on the surface following mining. Reclamation can be done periodically or contemporaneously with mining operations.

Periodic reclamation may occur at area strip mines, where reclamation activities may follow the advancing strip pit by a specified distance. Alternatively, as may occur with a large open pit, reclamation may have to be deferred to the end of the mining operations where the pit expands with time or is needed to provide space for production and mining support traffic.

Much of the reclamation work is part of handling the overburden in the mining sequence, leaving little but final grading and planting operations for postmining activities. With good reclamation planning, it is possible to have competitive, productive mining operations and unspoiled postmining landscapes at the same time as well as town infrastructure.

Successful reclamation is generally defined by regulatory standards, private or community agreement, or the goals of the mining company. Surface mining requires handling the entire surface and overburden mass above the oil shale—reclaiming and restructuring this mass offers a greater opportunity to select and structure the land for the needs and purposes of the human and natural community than would be available in almost any other way. Selecting a postmining land use must be based on cultural factors in the surrounding community, subject to the constraints of the environment that are beyond human control.

The *choice of postmining land use* (including the type of land ownership) is often a significant factor in reclamation. Land leased from a farmer would often be reclaimed for agricultural use, whereas government-held land in recent years has often been used for parks, for recreation, or as wildlife habitat. Surrounding land use should be considered since land may often be reclaimed to its former use or to the predominant land use in the immediate area.

7.5 PRODUCTION OF SHALE OIL

After mining, the oil shale is transported to a facility for retorting, after which the oil must be upgraded by further processing before it can be sent to a refinery, and the spent shale must be disposed of, often by putting it back into the mine. Eventually, the mined land is reclaimed. Both mining and processing of oil shale involve a variety of environmental impacts, such as global warming and greenhouse gas emissions, disturbance of mined land, disposal of spent shale, use of water resources, and impacts on air and water quality. The development of a commercial oil shale industry in the United States would also have significant social and economic impacts on local communities. Other impediments to development of the oil shale industry in the United States include the relatively high cost of producing oil from oil shale (currently greater than $60 per barrel) and the lack of regulations over leasing oil shale.

There is the potential for generating significant levels of atmospheric pollutants from every major operation in an oil-shale-retorting facility. These pollutants include coal dust, combustion products, fugitive organics, and fugitive gases. The fugitive organics and gases could include carcinogenic polynuclear organics and toxic gases such as oxides of carbon, hydrogen sulfide, ammonia, sulfur oxides, and mercury.

Emissions from auxiliary processes include combustion products from on-site steam/electric power plants and volatile emissions from wastewater systems, cooling towers, and fugitive emission sources. Volatile emissions from cooling towers, wastewater systems, and fugitive emission sources possibly can include every chemical compound present in the retorting process. Compounds that can be present include hazardous organics; metal carbonyls; trace elements such as mercury; and toxic gases such as carbon dioxide, hydrogen sulfide, hydrogen cyanide, ammonia, carbonyl sulfide, and carbon disulfide.

Environmental factors, including the potential quantity and the composition of the pollutants in addition to those compounds in the fuel stream recognized as mutagens that are initially present in internal process and utility streams, are a necessary consideration in the operation of fossil fuel conversion plants.

In terms of innovative technologies that might be applied for the cleaner production of shale oil, both conventional and in situ retorting processes result in inefficiencies that reduce the volume and quality of the produced shale oil. Depending on the efficiency of the process, a portion of the kerogen that does not yield any liquid is either deposited as coke on the host mineral matter or converted to hydrocarbon gases. For the purpose of producing shale oil, the optimal process is one that minimizes the regressive thermal and chemical reactions that form coke and hydrocarbon gases and maximizes the production of shale oil. Novel and advanced retorting and upgrading processes seek to modify the processing chemistry to improve recovery and/or create high-value by-products. Novel processes are being researched and tested in laboratory-scale environments. Some of these approaches (like those advocated for coal processing) (Speight, 2013) include lower heating temperatures, higher heating rates, shorter residence time durations, introduction of scavengers such as hydrogen (or hydrogen transfer/donor agents), and introduction of solvents.

REFERENCES

Berkowitz, N., 1985. The Chemistry of Coal. Elsevier, Amsterdam, Netherlands.

Kahn, H., 1979. Toxicity of oil shale chemical products. A Review. Scand. J. Work Environ. Health 5 (1), 1–9.

Speight, J.G., 2007. The Chemistry and Technology of Petroleum, fourth ed. CRC Press, Boca Raton, Florida.

Speight, J.G., 2008. Synthetic Fuels Handbook: Properties, Processes, and Performance. McGraw-Hill, New York, NJ.

Speight, J.G., 2013. The Chemistry and Technology of Coal, third ed. CRC-Taylor and Francis Group, Boca Raton, FL.

US OTA, 1980. An Assessment of Oil Shale Technologies, Volume I. Report PB80-210115. Office of Technology Assessment. Congress of the United States, Washington, DC.

US DOE, 2004a. Strategic Significance of America's Oil Shale Reserves, I. Assessment of Strategic Issues, March. http://www.fe.doe.gov/programs/reserves/publications

US DOE, 2004b. Strategic Significance of America's Oil Shale Reserves, II. Oil Shale Resources, Technology, and Economics; March. http://www.fe.doe.gov/programs/reserves/publications

US DOE, 2004c. America's Oil Shale: A Roadmap for Federal Decision Making; USDOE Office of US Naval Petroleum and Oil Shale Reserves. http://www.fe.doe.gov/programs/reserves/publications

US DOE, September 2006. Fact Sheet: Oil Shale Water Resources. Office of Petroleum Reserves, US Department of Energy, Washington, DC.

Alcohol: The family name of a group of organic chemical compounds composed of carbon, hydrogen, and oxygen. The molecules in the series vary in chain length and are composed of a hydrocarbon plus a hydroxyl group. Alcohols include methanol and ethanol.

Alkylation: A process for manufacturing high octane blending components used in unleaded petrol or gasoline.

API gravity: A measure of the lightness or heaviness of petroleum that is related to density and specific gravity.

$$°API = (141.5/sp\ gr\ @\ 60°F)\ 131.5.$$

Aromatics: A range of hydrocarbons that have a distinctive sweet smell and include benzene and toluene, which occur naturally in petroleum. They are extracted as a petrochemical feedstock as well as for use as solvents.

Asphaltene fraction: The brown to black powdery material produced by treatment of petroleum, heavy oil, bitumen, or residuum with a low-boiling liquid hydrocarbon.

Barrel (bbl): The unit of measure used by the petroleum industry; equivalent to approximately 42 US gallons or approximately thirty four (33.6) Imperial gallons or 159 liters; 7.2 barrels are equivalent to one tonne of oil (metric).

Barrel of oil equivalent (boe): The amount of energy contained in a barrel of crude oil, i.e., approximately 6.1 GJ (5.8 million Btu), equivalent to 1,700 kWh.

Billion: 1×10^9.

Bitumen: The extractable material in oil shale; also, on occasion, referred to as native asphalt and extra heavy oil; more correctly—a naturally occurring material that has little or no mobility under reservoir conditions and which cannot be recovered through a well by conventional oil well production methods, including currently used enhanced recovery techniques; successful methods involve mining for bitumen recovery.

British thermal unit (Btu): A nonmetric unit of heat, still widely used by engineers; one Btu is the heat energy needed to raise the temperature of one pound of water from 60°F to 61°F at one atmosphere pressure. 1 Btu = 1,055 joules (1.055 kJ).

Carbon dioxide (CO_2): A product of combustion that acts as a greenhouse gas in the Earth's atmosphere, trapping heat and contributing to climate change.

Carbon monoxide (CO): A lethal gas produced by incomplete combustion of carbon-containing fuels in internal combustion engines. It is colorless, odorless, and tasteless. (As in flavorless, we mean, though it's also been known to tell a bad joke or two.)

Catalyst: A substance that accelerates a chemical reaction without itself being affected. In refining, catalysts are used in the cracking process to produce blending components for fuels.

Coking: A thermal method used in refineries for the conversion of bitumen and residua to volatile products and coke (see Delayed coking and Fluid coking).

Conventional crude oil (conventional petroleum): Crude oil that is pumped from the ground and recovered using the energy inherent in the reservoir; also recoverable by application of secondary recovery techniques.

Cracking: A secondary refining process that uses heat and/or a catalyst to break down high molecular weight chemical components into lower molecular weight products, which can be used as blending components for fuels.

Delayed coking: A coking process in which the thermal reactions are allowed to the proceed to completion to produce gaseous, liquid, and solid (coke) products.

Density: The mass (or weight) of a unit volume of any substance at a specified temperature; see also Specific gravity.

Desulfurization: The removal of sulfur or sulfur compounds from a feedstock.

Diesel fuel: A distillate of fuel oil that has been historically derived from petroleum for use in internal combustion engines, also derived from plant and animal sources.

Distillate: Any petroleum product produced by boiling crude oil and collecting the vapors produced as a condensate in a separate vessel, for example gasoline (light distillate), gas oil (middle distillate), or fuel oil (heavy distillate).

Distillation: The primary distillation process that uses high temperature to separate crude oil into vapor and fluids which can then be fed into a distillation or fractionating tower.

Effluent: The liquid or gas discharged from a process or chemical reactor, usually containing residues from that process.

Emissions: Waste substances discharged into the air, land, or water.

Energy-efficiency ratio: A number representing the energy stored in a fuel as compared to the energy required to produce, process, transport, and distribute that fuel.

Fluid coking: A continuous fluidized solids process that cracks feed thermally over heated coke particles in a reactor vessel to gas, liquid products, and coke.

Fluidized-bed boiler: A large, refractory-lined vessel with an air distribution member or plate in the bottom, a hot gas outlet in or near the top, and some provisions for introducing fuel; the fluidized bed is formed by blowing air up through a layer of inert particles (such as sand or limestone) at a rate that causes the particles to go into suspension and continuous motion.

Fossil fuel: Solid, liquid, or gaseous fuels formed in the ground after millions of years by chemical and physical changes in plant and animal residues under high temperature and pressure. Oil, natural gas, and coal are fossil fuels.

Fuel oil: A heavy residue, black in color, used to generate power or heat by burning in furnaces.

Greenhouse gases: Gases that trap the heat of the sun in the Earth's atmosphere, producing the greenhouse effect. The two major greenhouse gases are water vapor and carbon dioxide. Other greenhouse gases include methane, ozone, chlorofluorocarbons, and nitrous oxide.

Heating value: The maximum amount of energy that is available from burning a substance.

Heavy oil (heavy crude crude): Oil that is more viscous than conventional crude oil, has a lower mobility in the reservoir but can be recovered through a well from the reservoir by the application of a secondary or enhanced recovery methods.

Heteroatomic compounds: Chemical compounds that contain nitrogen and/or oxygen and/or sulfur and/or metals bound within their molecular structure(s).

Hydrocarbonaceous material: A material such as bitumen that is composed of carbon and hydrogen with other elements (heteroelements) such as nitrogen, oxygen, sulfur, and metals chemically combined within the structures of the constituents; even though carbon and hydrogen may be the predominant elements, there may be very few true hydrocarbons (q.v.).

Hydrodesulfurization: The removal of sulfur by hydrotreating.

Hydroprocesses: Refinery processes designed to add hydrogen to various products of refining.

Hydrotreating: The removal of heteroatomic (nitrogen, oxygen, and sulfur) species by treatment of a feedstock or product at relatively low temperatures in the presence of hydrogen.

Inclined grate: A type of furnace in which fuel enters at the top part of a grate in a continuous ribbon, passes over the upper drying section where moisture is removed, and descends into the lower burning section. Ash is removed at the lower part of the grate.

Kerosene: A light middle distillate that in various forms is used as aviation turbine fuel or for burning in heating boilers or as a solvent, such as white spirit.

Methanol: A fuel typically derived from natural gas, but which can be produced from the fermentation of sugars in biomass.

Million: 1×10^6.

Moisture content: The weight of the water contained in a fossil fuel, usually expressed as a percentage of weight, either oven-dry or as received.

Nitrogen oxides (NOx): Products of combustion that contribute to the formation of smog and ozone.

Oil from tar sand: Synthetic crude oil (q.v.).

Oil mining: Application of a mining method to the recovery of bitumen.

Pay zone thickness: The depth of a oil shale deposit from which shale oil can be produced and recovered.

Particulate: A small, discrete mass of solid or liquid matter that remains individually dispersed in gas or liquid emissions.

Particulate emissions: Particles of a solid or liquid suspended in a gas, or the fine particles of carbonaceous soot and other organic molecules discharged into the air during combustion.

Process heat: Heat used in an industrial process rather than for space heating or other housekeeping purposes.

Pyrolysis: The thermal decomposition of biomass at high temperatures (greater than 200°C or 400°F) in the absence of air; the end product of pyrolysis is a mixture of solids (char), liquids (oxygenated oils), and gases (methane, carbon monoxide, and carbon dioxide) with proportions determined by operating temperature, pressure, oxygen content, and other conditions.

Refractory lining: A lining, usually of ceramic, capable of resisting and maintaining high temperatures.

Residuum (pl. residua, also known as resid or resids): The non-volatile portion of petroleum that remains as residue after refinery distillation; hence, atmospheric residuum and vacuum residuum.

Sandstone: A sedimentary rock formed by compaction and cementation of sand grains, which can be classified according to the mineral composition of the sand and cement.

Specific gravity: The mass (or weight) of a unit volume of any substance at a specified temperature compared to the mass of an equal volume of pure water at a standard temperature.

Synthetic crude oil (syncrude): A hydrocarbon product produced by the conversion of coal, oil shale, or tar sand bitumen that resembles conventional crude oil; which can be refined in a petroleum refinery.

Tar sand (bituminous sand): A formation in which the bituminous material (bitumen) is found as a filling in veins and fissures in fractured rocks or impregnating relatively shallow sand, sandstone, and limestone strata. A sandstone reservoir that is impregnated with a heavy, extremely viscous, black hydrocarbonaceous, petroleum-like material that cannot be retrieved through a well by conventional or enhanced oil recovery techniques; the several rock types that contain an extremely viscous hydrocarbon which is not recoverable in its natural state by conventional oil well production methods including currently used enhanced recovery techniques.

Ton (US ton): 2,000 pounds.

Tonne (Imperial ton, long ton, shipping ton): 2,240 pounds; equivalent to 1,000 kilograms or in crude oil terms about 7.5 barrels of oil.

Traveling grate: A type of furnace in which assembled links of grates are joined together in a perpetual belt arrangement. Fuel is fed in at one end and ash is discharged at the other.

Trillion: 1×10^{12}.

Vacuum distillation: A secondary distillation process which uses a partial vacuum to lower the boiling point of residues from primary distillation and extract further blending components.

Viscosity: A measure of the ability of a liquid to flow or a measure of its resistance to flow; the force required to move a plane surface of area 1 square meter over another parallel plane surface 1 meter away at a rate of 1 meter per second when both surfaces are immersed in the fluid; the higher the viscosity, the slower the liquid flows.

Waste streams: Unused solid or liquid by-products of a process.

Printed and bound by CPI Group (UK) Ltd, Croydon, CR0 4YY

03/10/2024

01040423-0019